Pyroelectric Infrared Detectors

Stephen G. Porter BSc, PhD, MInstP

Pyroelectric Infrared Detectors
ISBN 978-1-3999-1070-5

Published by:
S. G. Porter,
Towcester,
UK

Table of Contents

1. Introduction..1

2. Pyroelectricity...2

    2.1.   History...2

    2.2.   The pyroelectric Effect...2

        2.2.1.   Crystal Structure and Ionic Polarization....................................3

        2.2.2.   Thermodynamic Modelling..5

3. The Pyroelectric Detector..14

    3.1.   Introduction..14

    3.2.   Basic Concepts...14

    3.3.   Amplifiers...17

    3.4.   Structures...20

    3.5.   Basic Device Physics...23

        3.5.1.   Black Body Radiation..23

        3.5.2.   Infrared Absorption...25

        3.5.3.   Thermal Structure..27

    3.6.   Responsivity...27

        3.6.1.   Current Responsivity...28

        3.6.2.   Voltage Responsivity..31

    3.7.   Noise..36

        3.7.1.   Thermal Noise...37

        3.7.2.   Resistor Noise...38

        3.7.3.   Dielectric Noise...38

        3.7.4.   Amplifier Noise...39

        3.7.5.   Total Noise..39

3.7.6. Non-Gaussian Noise..40

3.8. Figures of Merit 41

3.8.1. Noise Equivalent Power 41

3.8.2. Noise Equivalent Irradiance 41

3.8.3. Detectivity 42

3.9. Performance Examples..42

3.10. Microphony 45

4. Pyroelectric Materials..46

4.1. Triglycine Sulphate 47

4.2. Lithium Tantalate...48

4.3. Strontium Barium Niobate 48

4.4. Polyvinylidene Fluoride...48

4.5. PZT Pyroelectric Ceramics 49

4.6. Comparative Performance..50

5. Infrared Detection Devices 53

5.1. Introduction..53

5.2. Single Element Detectors...53

5.2.1. Alternative Bias Arrangements 54

5.2.2. Edge electrodes 59

5.3. Compensated Detectors..60

5.3.1. Parallel Compensation...61

5.3.2. Series Compensation 65

5.4. Dual and Quad Element Detectors 70

5.5. Multi-element Arrays...74

5.5.1. Linear arrays..74

5.5.2. Two dimensional arrays 75

5.5.3. Read-Out Integrated Circuits..78

iv

5.6. The Pyroelectric Vidicon 79

6. Applications of Infrared Detectors...82

   6.1. Introduction...82

   6.2. Sensors 82

      6.2.1. Intruder Alarms 82

      6.2.2. Light Switches and Occupancy Sensors.........................87

      6.2.3. People Counters..88

      6.2.4. Flame Detectors..89

   6.3. Monitors and Radiometers 90

      6.3.1. Infrared Spectral Analysis 90

      6.3.2. Laser Monitors 91

      6.3.3. Radiometers...94

   6.4. Thermal Imaging..96

      6.4.1. Modulation 97

      6.4.2. Image processing..100

      6.4.3. Thermography 102

   6.5. Comparison with Alternative Detectors 104

      6.5.1. Thermopiles...104

      6.5.2. Resistive bolometers 104

      6.5.3. Photo-voltaic detectors 105

7. Infrared Optics 106

   7.1. Infrared Windows 106

      7.1.1. Germanium Windows..106

      7.1.2. Silicon Windows 109

      7.1.3. Other Window Materials 109

   7.2. Infrared Lenses...109

      7.2.1. Plastic Lenses 109

7.2.2.   Germanium Lenses..109

7.2.3.   Other Lens Materials..110

7.3.   Mirrors and Distorting Optics 110

8.   Conclusion 111

9.   References...112

Acknowledgements 115

About the Author..116

1. Introduction

After dusk a few weeks ago, I was walking along a street where I live when I noticed that, as I approached each house, a light switched on outside the front door. Why and how did this happen? The answer to this question is that each house had an external light fitted with a passive infrared light switch, probably incorporating a pyroelectric detector.

The aim of this little book is to explain some of the basic physics behind pyroelectric detectors and to describe something of the design, operation, and applications of pyroelectric devices. We will start with an introduction to pyroelectricity.

2. Pyroelectricity

2.1. History

The strange effects occurring when the mineral tourmaline is heated have been known for many hundreds of years [1]. It is generally thought that tourmaline is being referred to when the Greek philosopher Theophrastus refers to 'lyngourion'. A natural history book, *Hortus Sanitatis Major*, printed in 1497, states [1] [2]: "the gem Ligurius is called in this way because it comes from the solidified urine of the lynx. It is tawny, like amber and attracts leaves that come near it by energy. Theophrastus says that the ligurium has the colour of amber, that it attracts chaff, that it soothes aching stomachs, that it gives back colour to people affected by jaundice, and that it contracts the movement of the bowels."

The term pyroelectricity was first used by Brewster [3] in 1824. He studied the phenomenon in a variety of substances, including Rochelle salt. It was Rochelle salt in which ferroelectric properties were first observed by Valasek [4] in 1921.

The importance of the pyroelectric effect in infrared detection was emerging in 1970, and it was in this year that a widely acclaimed review of pyroelectric detectors by Putley was published [5]. At that time the only pyroelectric material of significant value was triglycine sulphate. A considerable amount of research and development was devoted to pyroelectric detectors in the latter quarter of the twentieth century, covering all aspects of materials, device fabrication and applications. An update of Putley's review appeared in 1977 [6], and the current author published a review in 1981 [7].

2.2. The pyroelectric Effect

A pyroelectric material is one which possesses an inherent electrical polarization, the magnitude of which is a function of temperature [3] [8] [9]. Most pyroelectrics are also ferroelectric, which means that the direction of their polarization can be reversed by the application of a suitable electric field, and their polarization reduces to zero at some temperature known as the Curie temperature, T_c, by analogy with ferromagnetism [10]. The dependence of polarization on temperature is typically of the form illustrated in Figure 2.1. The gradient of this curve dP/dT at a particular temperature T is the pyroelectric coefficient, p.

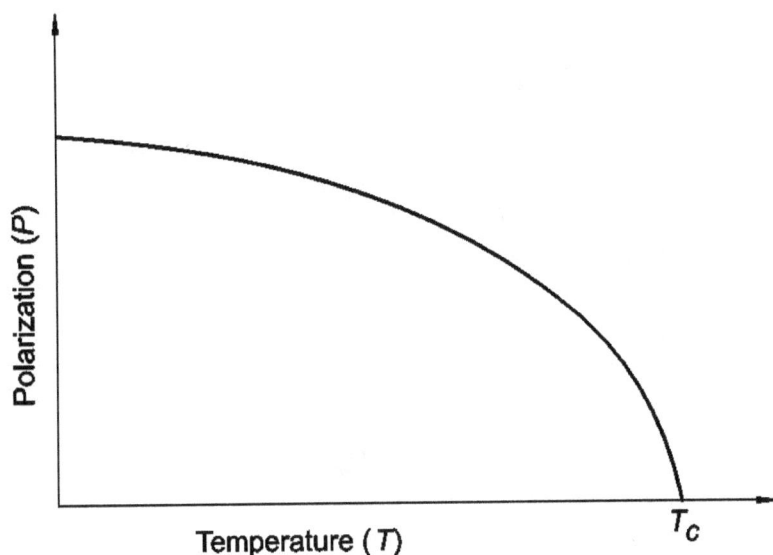

Figure 2.1 Temperature dependence of polarization for a ferroelectric material.

It has been said that all ferroelectrics are pyroelectric, but not all pyroelectrics are ferroelectric. However, the distinction between ferroelectric and pyroelectric materials given above is somewhat arbitrary because the ability to reverse the polarization of a given material may depend on experimental limitations such as crystal perfection, electrical conductivity, temperature, pressure, etc. An alternative distinction is that the polarization of a ferroelectric decreases to zero at some temperature, but it could be argued that some ferroelectrics decompose before the Curie temperature is reached. In general, no distinction is made between ferroelectricity and pyroelectricity in the theoretical treatment of the phenomena.

2.2.1. Crystal Structure and Ionic Polarization

Crystal structures may be divided into 32 crystal classes according to their symmetry. Of these, 21 classes do not have a centre of symmetry, and 10 of these have a unique polar axis and possess a polarisation which is generally a function of temperature. Crystals of this type are pyroelectric. For a comprehensive treatment of crystal structure see, for example, Zhdanov [11].

Consider the hypothetical two-dimensional crystal structure represented in Figure 2.2, in which the large circles with crosses represent positive ions and the smaller circles with dashes represent negative ions. This figure represents the configuration at temperatures above the Curie point, T_c. If the temperature is reduced below the Curie point, the equilibrium positions of the negative ions move with respect to the positive ions, as illustrated in Figure 2.3, and a spontaneous polarization is generated. If an external electric field is applied, the negative ions may be forced to move in the opposite direction relative to the positive ions, and the direction of polarization is reversed.

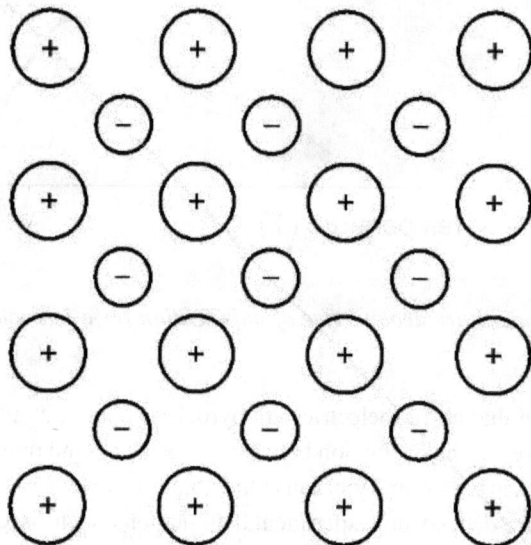

Figure 2.2 A hypothetical two-dimensional ionic crystal at T>T$_c$

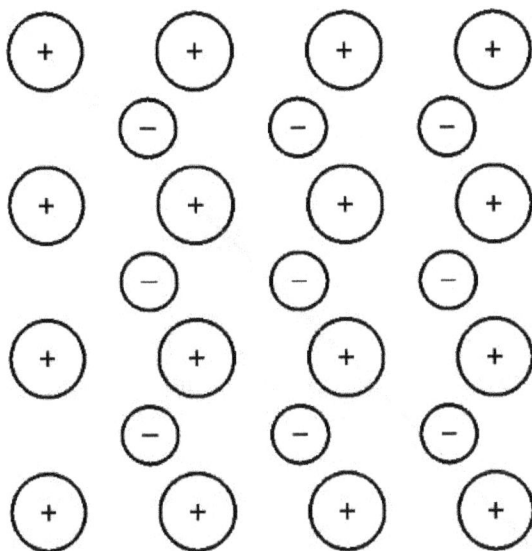

Figure 2.3 A hypothetical two-dimensional ionic crystal at T<T_c

2.2.2. Thermodynamic Modelling

2.2.2.1. Free Energy and Polarization

If we consider a ferroelectric material with one ferroelectric axis in the absence of any external stress, the free energy, G, relative to the un-polarized state, may be written as:

$$G = \frac{1}{2}aP^2 + \frac{1}{4}bP^4 + \frac{1}{6}cP^6 \qquad (2.1)$$

where P is the polarization and $a, b, and\ c$ are coefficients. The series is truncated at the third term, and only even powers are included because the energy must be the same for both directions of polarization, i.e. when P is negative or positive.

If an electric field, E, is applied the free energy will change in proportion to E and P (at least when E and P are close to zero). Equation 2.1 then becomes:

$$G = \frac{1}{2}aP^2 + \frac{1}{4}bP^4 + \frac{1}{6}cP^6 - EP \qquad (2.2)$$

and:

$$\frac{\delta G}{\delta P} = aP + bP^3 + cP^5 - E \tag{2.3}$$

The essence of the Landau-Devonshire theory [12] is that a phase transition occurs at a temperature, T_o, and the coefficient a changes sign at this temperature, and this is expressed as:

$$a = a_o(T - T_o) \tag{2.4}$$

The other coefficients are assumed to be independent of temperature. Note that T_o, the temperature at which a changes sign, is close to, but not necessarily equal to the Curie temperature, T_c. Substituting equation 2.4 into 2.1 and 2.3 gives:

$$G = \frac{1}{2}a_o(T - T_o)P^2 + \frac{1}{4}bP^4 + \frac{1}{6}cP^6 \tag{2.5}$$

and:

$$\frac{\delta G}{\delta P} = a_o(T - T_o)P + bP^3 + cP^5 - E \tag{2.6}$$

The spontaneous polarization, P_o, is the value of P when the electric field is zero and G is a minimum, so $\frac{\delta G}{\delta P}$ is zero. Putting $P = P_o$, $E = 0$, and $\frac{\delta G}{\delta P} = 0$ into equation 2.6 gives:

$$cP_o^5 + bP_o^3 + a_o(T - T_o)P_o = 0 \tag{2.7}$$

This has two possible solutions:

$$P_o = 0 \tag{2.8}$$

or:

$$cP_o^4 + bP_o^2 + a_o(T - T_o) = 0 \tag{2.9}$$

If b is positive, solving equation 2.9 as a quadratic equation in P_o^2, we find that it has real solutions for P_o only if $T < T_o$. So for temperatures above T_o the only solution is equation 2.8, i.e. $P_o = 0$, and for temperatures below T_o the solution to equation 2.9 gives P_o as a function of T of the form shown in Figure 2.4, and we have a second order phase transition.

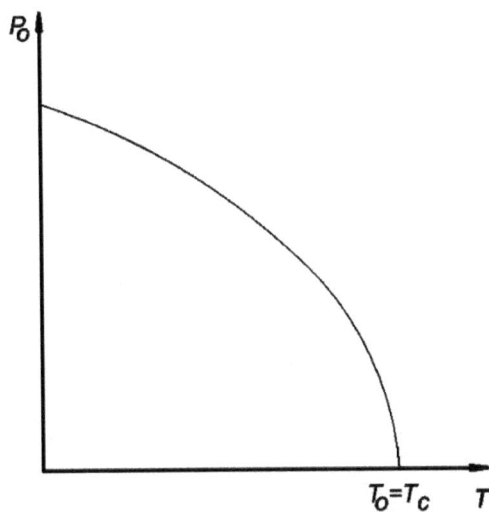

Figure 2.4 *Spontaneous polarization as a function of temperature for a second order phase transition.*

If b is negative, equation 2.9 has real solutions when $T < T_c$, where

$$T_c = T_o + \frac{b^2}{4a_o c} \qquad (2.10)$$

In this case we have a first order phase transition and P_o as a function of temperature has the form illustrated in Figure 2.5.

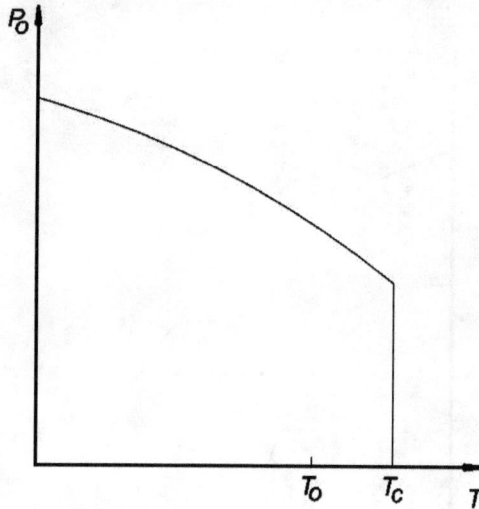

Figure 2.5 Spontaneous polarization as a function of temperature for a first order phase transition

From equation 2.5 we can plot the free energy as a function of polarization at different temperatures. Again, we get different results depending on whether b is positive or negative, as shown in Figure 2.6 and Figure 2.7.

We see that, in the case of a second order phase transition, at low temperatures there are two energy minima, corresponding to opposite directions of spontaneous polarization. As the temperature increases, these minima converge continuously until the transition temperature is reached, at which point they coalesce to form a single minimum and the spontaneous polarization decreases to zero. In this case the transition temperature, T_o, is equal to the Curie temperature, T_c.

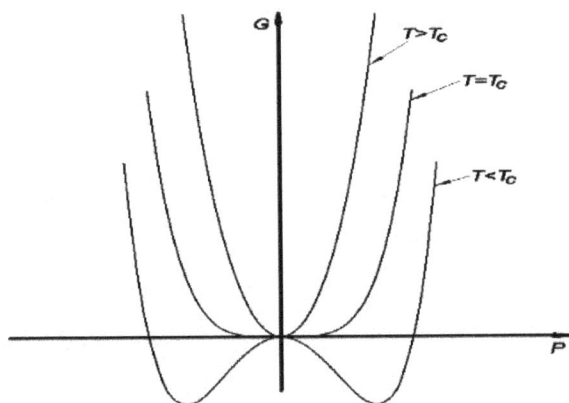

Figure 2.6 Free energy as a function of polarization for a second order phase transition.

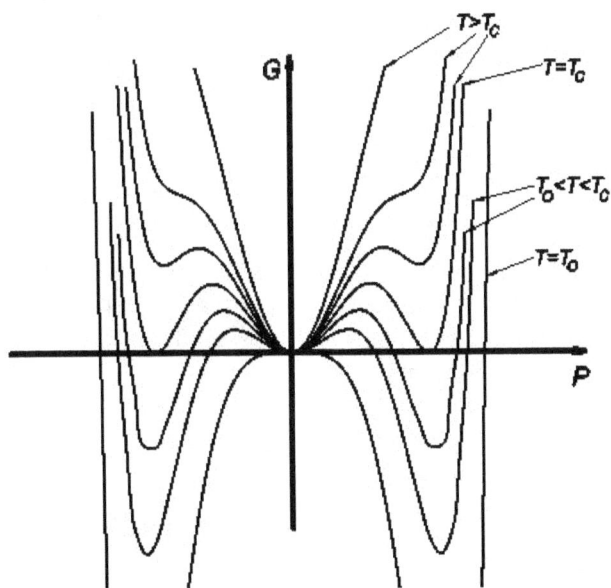

Figure 2.7 Free energy as a function of polarization for a first order phase transition.

In the case of a first order phase transition, there are also two energy minima at low temperatures, but, as the temperature increases, a third minimum appears at the origin and the three minima have equal values at the Curie temperature, and zero and finite polarization have equal probability. As the temperature moves away from the Curie temperature, the potential barrier between the zero and finite values of polarization decreases, and the polarization will switch to zero at high temperatures or a finite value at low temperatures. This results in hysteresis, with the switch occurring at a higher temperature when the temperature is increasing and a lower temperature when it is decreasing. In the case of a first order phase transition, T_o, the temperature at which the energy minimum at zero polarization disappears, is lower than the Curie temperature, T_c.

2.2.2.2. Domains

Normally, ferroelectric crystals are not uniformly polarised in one direction, but comprise a number of domains in which the directions of polarisation differ. For example, a tetragonal structure can have its polar axis oriented in any one of 6 mutually orthogonal directions. A typical tetragonal ferroelectric crystal would have a multitude of domains in which the polar axes are oriented randomly in these 6 directions in such a way that the whole crystal has zero net polarization.

This is illustrated in a simplified schematic manner for a hypothetical two-dimensional crystal in Figure 2.8. In two dimensions, there are four possible directions for the polar axis, and the illustration shows the crystal divided into domains such that the net polarization in any direction is zero.

In practice, in most cases, there will be many more domains of more random shapes and sizes, but the principle is the same. This means that, before a ferroelectric material can be used to generate charge, it must be polarized by the application of an electric field, usually at elevated temperature.

A polycrystalline ceramic material is made up of a multitude of small crystal grains oriented in random directions, and each of the grains will consist of several domains. When a polarizing electric field is applied, each grain will become a single domain with its polar axis aligned as close as possible to the direction of the applied field.

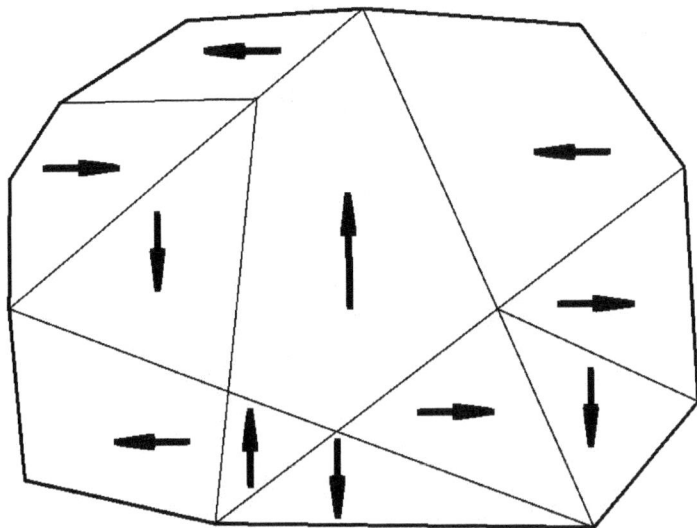

Figure 2.8 Simplified schematic domain structure.

2.2.2.3. Relative Permittivity

One more thing to consider is the behaviour of the relative permittivity near to the transition temperature. First, we note two equations defining the relationships between the dielectric susceptibility, χ, the polarisation, P, the electric field, E, and the relative permittivity, ε_r:

$$\chi = \lim_{E \to 0} \frac{\delta P}{\delta E} \qquad (2.11)$$

$$\varepsilon_r = 1 + \chi \qquad (2.12)$$

Taking equation 2.6 and, for simplicity, ignoring the P^5 term we have, when the energy is minimum and $\frac{\delta G}{\delta P} = 0$:

$$0 = a_o(T - T_o)P + bP^3 - E \qquad (2.13)$$

Differentiating with respect to E gives:

$$0 = a_o(T - T_o)\frac{\delta P}{\delta E} + 3bP^2\frac{\delta P}{\delta E} - 1 \tag{2.14}$$

and:

$$\frac{\delta P}{\delta E} = \frac{1}{a_o(T-T_o)+3bP^2} \tag{2.15}$$

As E tends towards zero, P approaches P_o and, from equation 2.13 putting $E = 0$ and $P = P_o$:

$$0 = a_o(T - T_o)P_o + bP_o^3 \tag{2.16}$$

so either:

$$P_o = 0 \tag{2.17}$$

or:

$$P_o^2 = \frac{a_o(T_o-T)}{b} \tag{2.18}$$

For temperatures above the transition temperature P_o is zero and, from equations 2.11, 2.15, and 2.17:

$$\chi = \frac{1}{a_o(T-T_o)} \tag{2.19}$$

and:

$$\varepsilon_r = 1 + \frac{1}{a_o(T-T_o)}, \quad T > T_o \tag{2.20}$$

For temperatures below the transition temperature, from 2.11, 2.15, and 2.18, we get:

$$\chi = \frac{1}{a_o(T-T_o)+3a_o(T_o-T)} \tag{2.21}$$

i.e.

$$\chi = \frac{1}{2a_o(T_o-T)} \tag{2.22}$$

and:

$$\varepsilon_r = 1 + \frac{1}{2a_o(T_o-T)}, \quad T < T_o \tag{2.23}$$

Using equations 2.20 and 2.23 we see that the relative permittivity as a function of temperature is of the form shown in Figure 2.9.

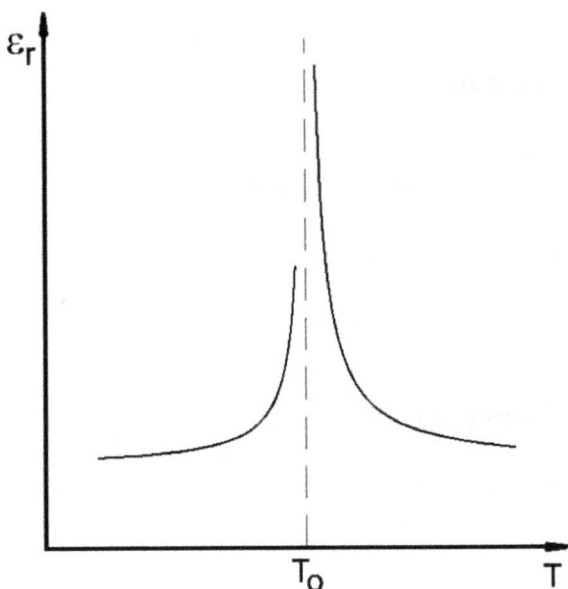

Figure 2.9 Relative permittivity as a function of temperature.

For more detailed treatment of the Landau-Devonshire theory of pyroelectrics, see, for example [10], [11], [12], [13], and [14].

3. The Pyroelectric Detector

3.1. Introduction

In this chapter we will discuss the structure and physics of pyroelectric infrared detectors. We cover the principles of thermal radiation and infrared absorption, and we will develop models of responsivity, noise, and figures of merit for infrared detectors.

The application of these principles to the design and construction of infrared detection devices we leave to Chapter 5, and the applications in which these devices may be used will be covered in Chapter 6.

3.2. Basic Concepts

A simple pyroelectric radiation detector element consists of a piece of pyroelectric material, of thickness d, with electrodes on opposite faces, the material being oriented such that its polar axis, c, is perpendicular to the electroded faces. This is shown schematically in Figure 3.1. As will be discussed later, in Chapter 5, there are a variety of configurations that may be used for the detector element electrodes. The simplest configuration, which will suffice for the present discussion, is an electrode which covers the full face A.

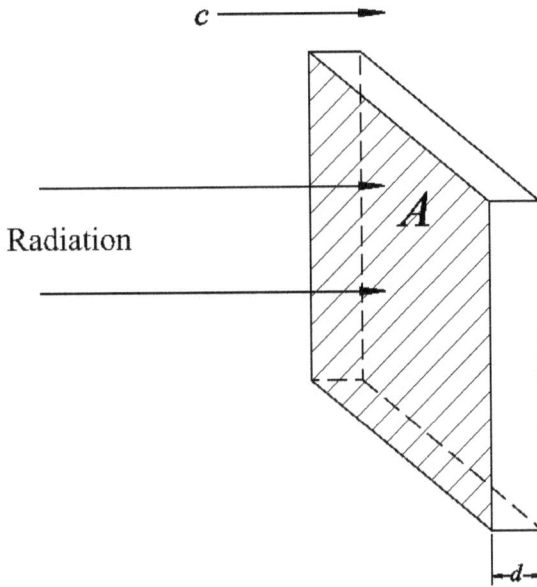

Figure 3.1 Schematic representation of a pyroelectric detector element, showing the polar axis, c, perpendicular to the electroded face, A.

For a pyroelectric detector to work, the energy of the incident radiation must be absorbed into the detector element as efficiently as possible to maximise the consequent change in temperature. To achieve this, the electrode on face *A* may be transparent or semi-transparent, allowing radiation to be absorbed in the pyroelectric material. Alternatively, the electrode may be absorbing or coated with a layer of material designed to absorb radiation at the wavelengths of greatest interest. This can be used to give a more uniform spectral response from the detector.

As noted in Chapter 2, a ferroelectric consists of many separate domains with differing directions of polarization, so that the net effect over the whole area is zero. Before use therefore, these domains must be re-oriented by the application of an electric field so that all become parallel to one another (or as near parallel as possible in the case of a ceramic). This is usually done at an elevated temperature so that the coercive field is reduced.

Even when fully poled, there will generally be no observable voltage across a pyroelectric detector element once thermal equilibrium with its surroundings has been established. This is because the internal polarization is balanced by a surface charge, which accumulates over time. It is only when the temperature of the detector is changed, generating a change in spontaneous polarization, that a signal current or voltage will be observed. For this reason, a pyroelectric detector can only be used to detect changes in irradiance (the intensity of radiation falling upon it).

When the detector is heated by the incident radiation, the polarization changes by an amount determined by the temperature change and the pyroelectric coefficient of the material. This change in polarization appears as a charge on the capacitor formed by the pyroelectric with its two electrodes, or as a current flowing in a circuit connected to them. If the incident radiation is modulated, for example by a mechanical chopper, an alternating current is generated, as illustrated schematically in Figure 3.2.

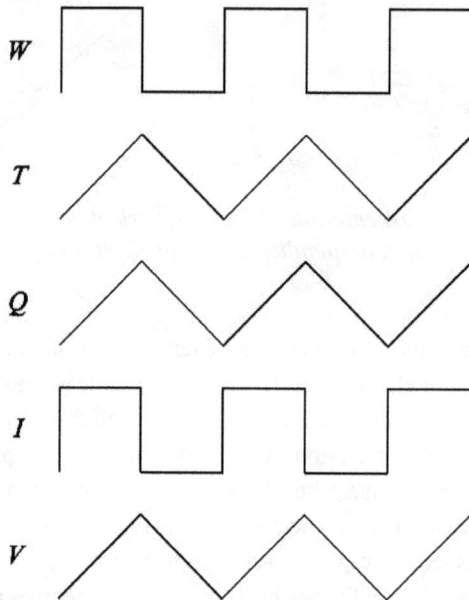

Figure 3.2   Schematic waveforms for a pyroelectric detector showing incident power, W, Temperature, T, pyroelectric charge, Q, current, I, and voltage, V.

Here W represents the incident power, T is the temperature of the pyroelectric, Q is the generated pyroelectric charge, I is the pyroelectric current, and V is the voltage on the capacitance of the detector.

The waveforms illustrated in Figure 3.2 are idealized for short timescales and small temperature excursions. As we will see later, these will be modified by the thermal and electrical time constants of the detector.

3.3. Amplifiers

Electronically, a pyroelectric detector may be considered as a current generator in parallel with a capacitor, as shown in Figure 3.3a. In this book we will use the symbol shown in Figure 3.3b to represent this combination, with the direction in which the chevron is pointing representing the direction of positive current flow when the pyroelectric is heated.

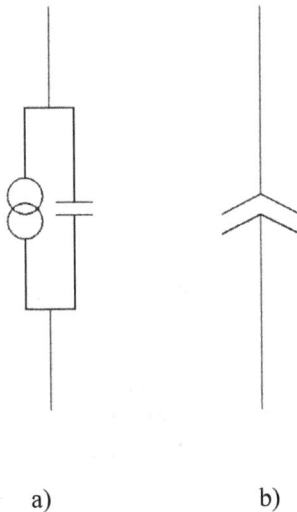

a)   b)

Figure 3.3 a) schematic diagram of a pyroelectric detector as a parallel combination of capacitor and current generator, and b) the symbol used to represent this throughout this book.

Typically, the pyroelectric current is of the order of 0.1 pA and the capacitance of the detector is of the order of 10 pF. In order to detect these very small charges, low noise high impedance amplifiers are required. The most common arrangement used in single element detectors is a simple junction field effect transistor (JFET) source follower, as shown in Figure 3.4. The bias resistor, R, is needed to provide the leakage current for the FET, and the load resistor, R_L, completes the source follower arrangement.

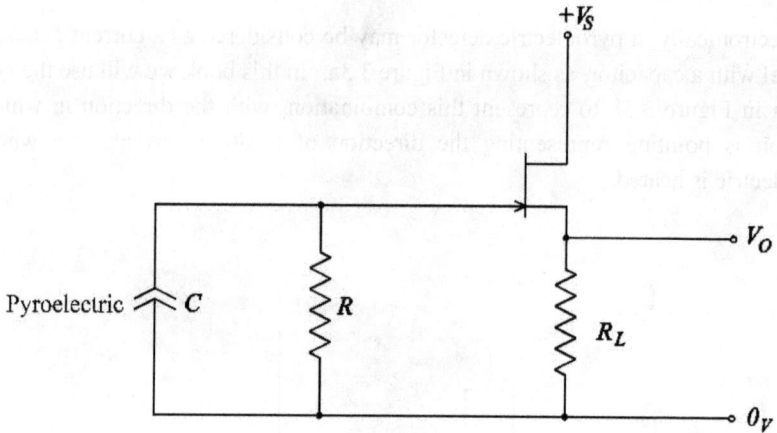

Figure 3.4 Schematic diagram of a pyroelectric detector with a JFET.

Special low leakage JFETs were developed for pyroelectric detectors during the latter half of the twentieth century. An example was the Texas Instruments BF800 which had a specified leakage current of less than 0.5 pA, giving a bias voltage of less than 0.25 V if R is 500 GΩ. Unfortunately, these devices are no longer available, and pyroelectric detector manufacturers tend to use low leakage selections of industry standard devices such as the 2N4117A or the MMBF4117, for which the specified leakage current is 1 pA.

As stated above, the capacitance, C, of a typical pyroelectric detector is of the order of 10 pF, so if R is 500 GΩ, the RC time constant is of the order of 5 seconds. In this case, for modulation frequencies greater than about 0.2 Hz, the output voltage, V_O, will be representative of the pyroelectric charge generated across the detector.

An alternative arrangement is to use a CMOS Operational Amplifier (Op Amp) instead of a JFET source follower. Figure 3.5 is the schematic diagram for a voltage amplifier, which generates an output voltage equal to the voltage across the pyroelectric multiplied by the gain, defined by the ratio R_1/R_2. Again, we have the lower frequency limit defined by the RC time constant, but the input leakage current of the CMOS amplifier may be lower than that of a JFET, so a higher value resistor may be used, giving a longer RC time constant.

Whether we use a JFET or a CMOS amplifier, there may be difficulty in meeting the requirements of both low input current noise and low voltage noise. This will be discussed later, when we have examined in more detail the parameters affecting the performance of a pyroelectric detector.

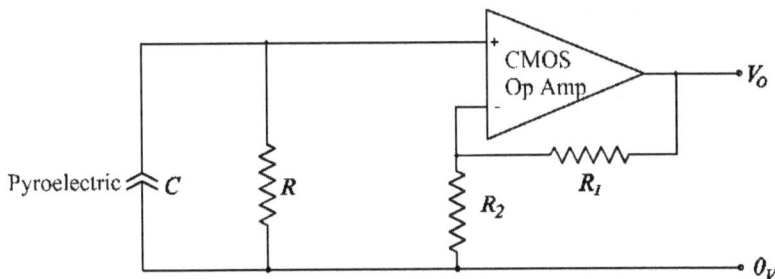

Figure 3.5 Schematic diagram of a pyroelectric detector with a CMOS voltage amplifier.

As noted above, the bias resistor, R, needs to be at least 100 GΩ to give a sufficiently long time constant to allow very low frequency signals to be processed. These high value resistors are expensive, and care is needed with their selection. Such resistors tend to have conduction mechanisms giving non-ohmic characteristics so that the actual resistance depends on the applied voltage, which in turn depends on the rate of change of temperature of the pyroelectric element.

An alternative arrangement using a CMOS operational amplifier is shown in Figure 3.6. This is a current-voltage converter, or trans-impedance amplifier (TIA) in which the current generated by the pyroelectric is made to flow through the feedback resistor, R, and the time constant is GRC, where G is the open loop gain of the amplifier. The value of R can therefore be reduced in proportion to G to achieve the same time constant.

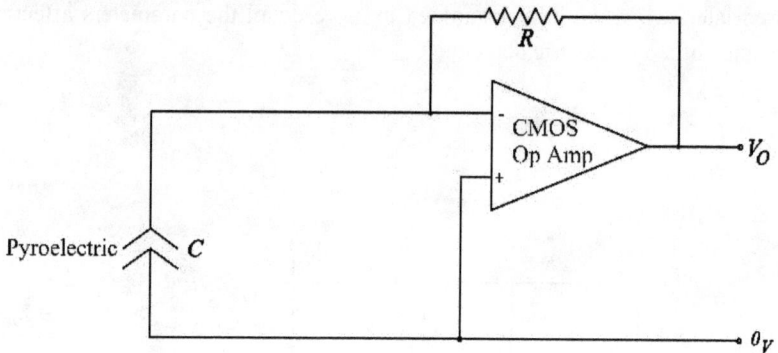

Figure 3.6 Schematic diagram of a pyroelectric detector with a CMOS current amplifier.

3.4. Structures

A typical single element pyroelectric detector is housed in a standard TO5 transistor package like the one shown in Figure 3.7. This has a metal lid into which is sealed an infrared transmitting window, typically made of germanium or silicon.

Figure 3.7 Photograph of a pyroelectric detector.

Figure 3.8 shows the structure inside such a detector. In this photograph, the pyroelectric element is a 2mm square of pyroelectric ceramic in the centre of the TO5 header. A JFET chip is bonded on top of the pyroelectric element with electrically conducting adhesive, and to its right is a 200 GΩ bias resistor. To the left of the pyroelectric, is the load resistor for the JFET source follower. Connections are made between the components using 25 μm diameter aluminium ultrasonic bonding wire. The gate of the JFET is connected both to the back surface of the chip and to a bond pad on the top surface, so this is connected to the pyroelectric by the conducting adhesive and to the bias resistor by a bond wire.

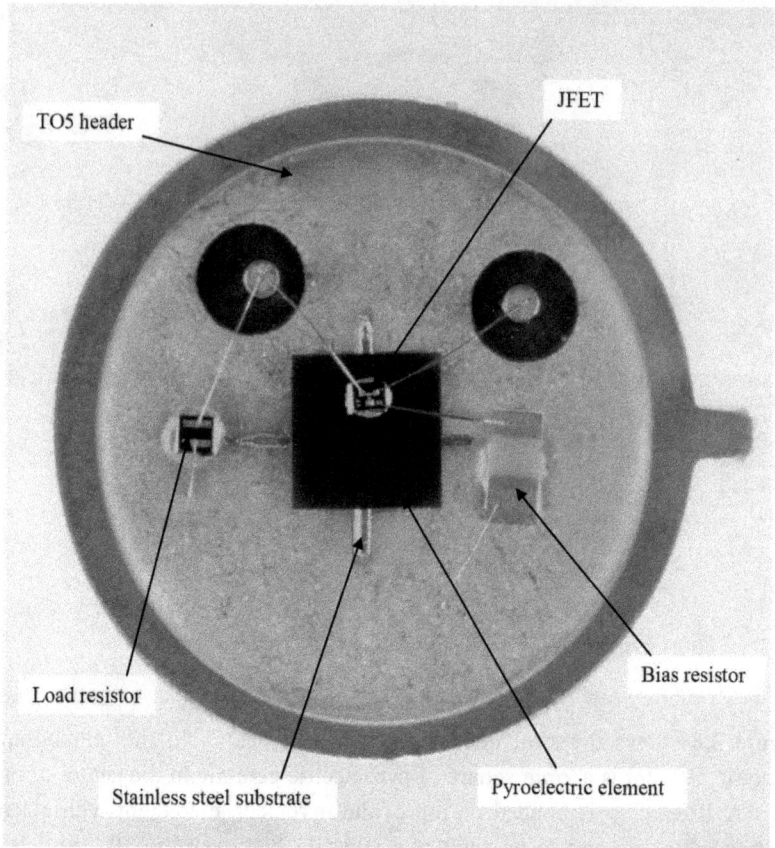

Figure 3.8 Photograph of the internal structure of a pyroelectric detector.

Underneath the pyroelectric chip is a stainless-steel substrate of cruciform shape, which supports the pyroelectric while allowing an air gap under most of the area between it and the header to provide good thermal isolation. Electrically conducting adhesive is used to hold these components in place and to provide electrical contact between the underside of the pyroelectric chip and the header surface, which is the electrical ground.

Although this structure lends itself to low-cost production techniques, there are some potential problems with mounting the FET chip directly on the pyroelectric. One is the possibility of thermal feedback, in which current through the FET dissipates heat which raises the temperature of the pyroelectric and generates a signal that increases the current in the FET. Another is the sensitivity of the FET to very small amounts of visible light leaking through the glass seals in the header or around the edges of the window. For this reason, some manufacturers use encapsulated FETs inside the package despite their potentially higher leakage and lower performance.

3.5. Basic Device Physics

As explained in section 3.2, a pyroelectric detector generates an output signal whenever its temperature is changed, for example, by absorbing some form of incident energy. In general, pyroelectric detectors are used to detect electromagnetic radiation, usually in the infrared region of the spectrum (although they have been used to sense other forms of energy input, such as the heat liberated in exothermic chemical reactions). Before considering the physics of the detector, we will therefore take a brief look at radiation theory.

3.5.1. Black Body Radiation

Any object radiates electromagnetic power as a function of wavelength and temperature according to Plank's law, which states that the spectral radiance is given by the formula

$$R_\lambda(T) = \frac{2hc^2\eta}{\lambda^5(e^{\frac{hc}{\lambda kT}}-1)} \tag{3.1}$$

where $R_\lambda(T)$ is the electromagnetic power emitted per unit solid angle at the wavelength λ for the object at absolute temperature T; h is Plank's constant, c is the speed of light, and k is Boltzmann's constant. η is the emissivity of the object at the wavelength λ, which, in the case of a black body, is equal to 1.

Putting $\eta = 1$ in equation 3.1 and calculating the spectral radiance of a black body at temperatures in the range -40 °C to +60 °C yields the results shown in Figure 3.9. Here the units of spectral radiation density are W sr^{-1} m^{-2} μm^{-1}. From this graph, we see that there is a peak in radiation at wavelengths in the range 8 μm to 12 μm for objects within the global ambient temperature range.

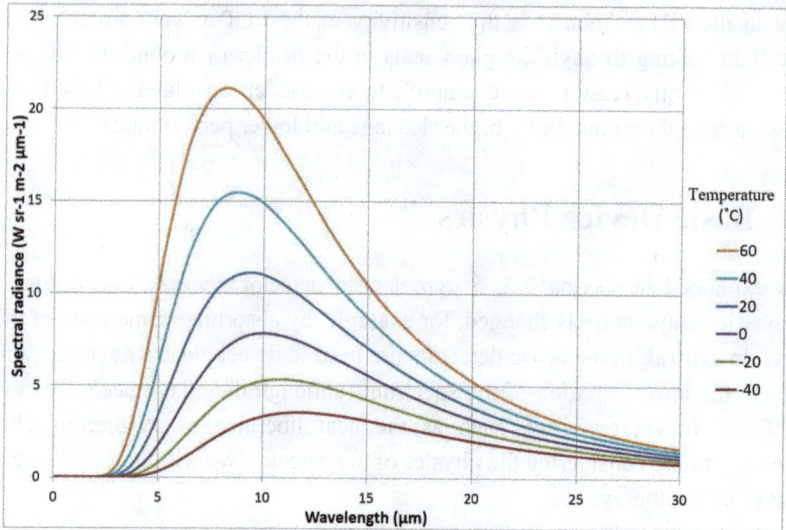

Figure 3.9 Spectral radiance of a black body at temperatures within the range -40 °C to +60 °C.

Thus, all objects radiate electromagnetic energy according to their temperature. For objects in the temperature range -170 °C to +500 °C the peak of this radiation lies within the mid-infrared wavelength band, 3 μm to 30 μm, and for objects at normal ambient temperature (-50 °C to +60 °C) the peak is in what is known as the thermal infrared band, 8 μm to 12 μm.

When designing infrared sensors, consideration must also be given to the transmission of radiation through the atmosphere. For example, over a path length of 2km, the atmosphere is opaque to wavelengths between 5.5 μm and 7.5 μm, mainly due to absorption by water vapour. There is, however, a broad transmission window between 8 μm and 14 μm, as shown in Figure 3.10.

Infrared detection by pyroelectric detectors is therefore generally concentrated within the thermal infrared wavelength band of 8 μm to 12 μm. There are, however, some specialist applications in which pyroelectric detectors are used in other wavebands, for example for the detection of specific gases by means of their absorption characteristics within the atmospheric window between 3μm and 5μm. These applications will be discussed in more detail in Chapter 6.

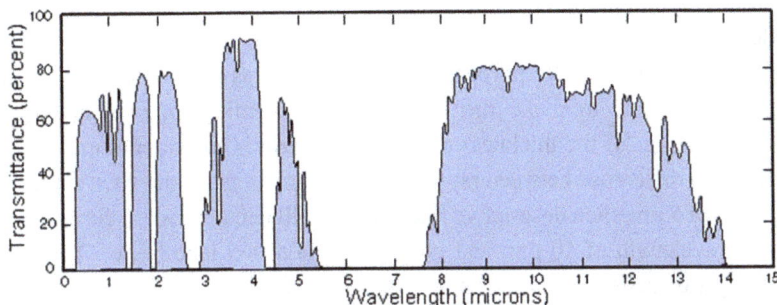

Figure 3.10 Atmospheric transmittance over one nautical mile as a function of wavelength. [15] [1]

3.5.2. Infrared Absorption

A pyroelectric detector will detect electromagnetic radiation at any wavelength that it can be made to absorb, and pyroelectric detectors have been used for millimetre waves (by making the pyroelectric 3 mm thick) and for x-rays. However, for wavelengths outside the mid-infrared band there are usually better, more sensitive detectors that can be used. In fact, the vast majority of pyroelectric detectors are used within the thermal band from 8 μm to 12 μm.

Thus, the first requirement of a thermal infrared detector is that it absorbs as much of the incident radiation as possible. The aim, therefore, is to make the effective emissivity, η, of the surface (which is equal to its absorptivity) as high as possible, while maintaining the thermal mass of the detector as low as possible, so that the temperature increase due to the absorbed radiation is as high as possible.

The simplest arrangement for a pyroelectric detector is a thin chip of pyroelectric material with metal electrodes on the two opposing major faces. If the pyroelectric material itself is a good absorber of 10μm radiation, then a transparent electrode, such as tin oxide, can be used on the surface exposed to the incident radiation, with a reflecting electrode, such as gold, on the back surface. Absorption then takes place within the pyroelectric, and the path length for absorption is twice the thickness of the

---

[1] This file is a work of a sailor or employee of the U.S. Navy, taken or made as part of that person's official duties. As a work of the U.S. federal government, the image is in the public domain.

chip, because any radiation that reaches the back surface is reflected by the back electrode.

A variation on this, if the pyroelectric is very thin, as may be the case in a two-dimensional array, is to make the front electrode semi-transparent and the back electrode reflecting. If the thickness of the pyroelectric is an odd multiple of 2.5 μm, so that the path difference between rays reflected from the front and back surfaces is an odd multiple of 5 μm, then destructive interference will occur between the two reflected rays for a wavelength of 10 μm, and good absorption will take place. There will be phase changes on reflection of rays at both surfaces, but as these will both be equal to π, they add up to 2π and therefore have no effect on the interference between the two reflected rays. In principle, the limitation of this is that the absorption is good only at the single wavelength of 10 μm, and only for normal incidence; however, there is usually significant absorption within the material itself and the thin film interference simply improves the performance at the important   10 μm wavelength.

A more common variation is to use a semi-transparent electrode, such as a thin layer of nichrome (an alloy of nickel and chromium) on the front surface of a relatively thick piece of pyroelectric. Any medium in which an electromagnetic wave is propagating has an intrinsic impedance given by [16]:

$$Z = \left(\frac{\mu_r \mu_0}{\varepsilon_r \varepsilon_0}\right)^{1/2} \tag{3.2}$$

For free space this reduces to:

$$Z_0 = \left(\frac{\mu_0}{\varepsilon_0}\right)^{1/2} \tag{3.3}$$

which has the value 376.7 ohms. (For a more detailed discussion of this, see any good textbook on Electricity and Magnetism, for example Bleaney and Bleaney [17]).

If a plane wave propagating in free space is incident on the surface of a medium whose impedance is Z_1, the reflection coefficient of the surface is given by:

$$B = \frac{Z_1 - Z_0}{Z_1 + Z_0} \tag{3.4}$$

For a metal surface, the modulus of Z_1 will be equal to ρ/δ, where ρ is the resistivity and δ is the skin depth, and for a very thin layer of metal on an insulating surface, δ can be taken as the thickness of the metal, and Z_1 becomes the surface resistance of the metal film in ohms per square.

So, if the thickness of a nichrome electrode is chosen such that the surface resistance is 376.7 Ω per square, expression 3.4 becomes equal to zero, and reflections at this first surface are minimised. This means that virtually all the incident radiation passes into the pyroelectric, where it can be absorbed.

Many pyroelectric materials, however, are not good absorbers at 10 μm, so an alternative arrangement is required. The solution is generally to apply a black, absorbing coating to the front surface of the detector chip, but this must be done without adding significantly to the thermal mass of the detector. Black paint, gold black, and candle black have all been used for this, but one of the best materials for this is electrochemically deposited platinum black [18]. This has been shown to absorb 98 % of incident radiation over the range 2 μm to 20 μm while adding a thermal capacity of only 1 J m^{-2} K^{-1}.

3.5.3. Thermal Structure

The pyroelectric detector chip has a thermal capacity, H, determined by its specific heat and its mass. As the detector chip must be mounted in some form of encapsulation, which acts as a heat sink, there is also a thermal conductance, G_T, to this heat sink. This means that the pyroelectric chip has a thermal time constant, H/G_T, as well as the electrical time constant, RC, mentioned in section 3.3.

3.6.   Responsivity

Let us consider a pyroelectric detector exposed to radiation which is modulated sinusoidally with amplitude W_0 at an angular frequency ω. The incident power can be expressed as a function of time, t, by:

$$W = W_0 e^{j\omega t} \qquad (3.5)$$

The temperature of the detector will be modulated by an amount which depends on the fraction of the incident radiation absorbed (i.e. the emissivity of the surface), η, the thermal capacity, H, and the thermal conductance, G_T, to the encapsulating heat sink. The temperature difference, T, between the detector and the heat sink is related to W by the differential equation:

$$\eta W = H \frac{dT}{dt} + G_T T \qquad (3.6)$$

i.e.

$$\eta W_0 e^{j\omega t} = H \frac{dT}{dt} + G_T T \qquad (3.7)$$

which has the solution:

$$T = \frac{\eta W_0 e^{j\omega t}}{\left(G_T^2 + \omega^2 H^2\right)^{1/2}} \qquad (3.8)$$

3.6.1. Current Responsivity

As explained in chapter 2, the polarization of a pyroelectric material changes with temperature. This change of internal polarization generates a surface charge given by:

$$\Delta Q = A \, \Delta P \qquad (3.9)$$

where A is the surface area and P is the polarization. If the temperature is changing with time, the pyroelectric current generated is:

$$i = \frac{dQ}{dt} = A \frac{dP}{dt} = A \frac{dP}{dT} \frac{dT}{dt} \qquad 3.10$$

The pyroelectric coefficient, p, is defined by:

$$p = \frac{dP}{dT} \qquad (3.11)$$

so we have:

$$i = Ap \frac{dT}{dt} \qquad (3.12)$$

Differentiating equation 3.8 and substituting into 3.12 gives:

$$i = \frac{j\eta p A W_0 \omega e^{j\omega t}}{\left(G_T^2 + \omega^2 H^2\right)^{1/2}} \qquad (3.13)$$

The current responsivity, $\mathfrak{R}_i$ is defined as the current generated by unit incident power, i.e.

$$\mathfrak{R}_i = \left| \frac{i}{W} \right| \qquad (3.14)$$

so we have, from 3.5, 3.13, and 3.14:

$$\mathfrak{R}_i = \left| \frac{j\eta p A\omega}{\left(G_T^2 + \omega^2 H^2\right)^{1/2}} \right| \tag{3.15}$$

i.e.

$$\mathfrak{R}_i = \frac{\eta p A\omega}{\left(G_T^2 + \omega^2 H^2\right)^{1/2}} \tag{3.16}$$

Putting equation 3.16 in terms of the thermal time constant, $\tau_T = \dfrac{H}{G_T}$, we get:

$$\mathfrak{R}_i = \frac{\eta p A\omega}{G_T\left(1 + \omega^2 \tau_T^2\right)^{1/2}} \tag{3.17}$$

In some cases, depending on the structure of the detector, the thermal conductance, G_T, is proportional to the detector area, and can be written as $G_T = gA$. Equation 3.17 then becomes:

$$\mathfrak{R}_i = \frac{\eta p\omega}{g\left(1 + \omega^2 \tau_T^2\right)^{1/2}} \tag{3.18}$$

Plotting $\mathfrak{R}_i$ as a function of ω gives the frequency dependence of the current responsivity, which takes the form shown in Figure 3.11.

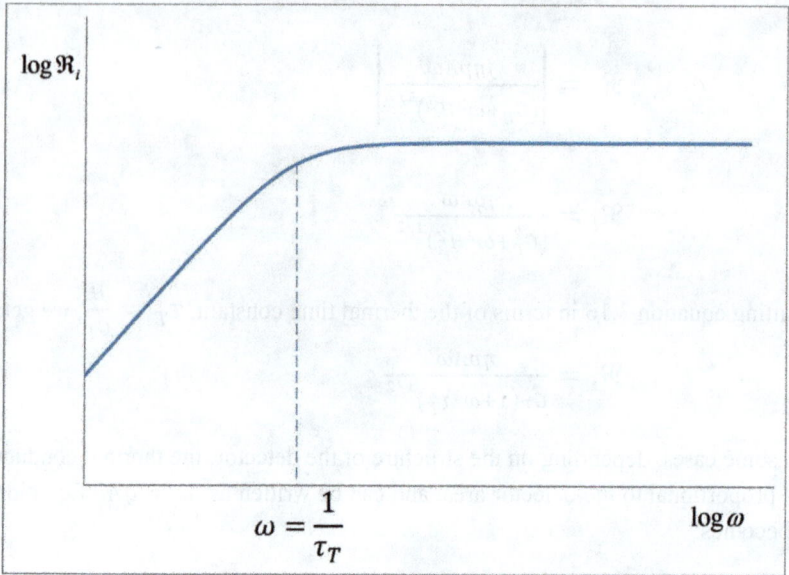

Figure 3.11 Current responsivity as a function of frequency.

For modulation frequencies that are high compared with the reciprocal of the thermal time constant, so that $\omega^2 \tau_T^2 \gg 1$, equation 3.17 reduces to:

$$\mathfrak{R}_i = \frac{\eta p A}{G_T \tau_T} \tag{3.19}$$

Substituting $\tau_T = H/G_T$ in this gives:

$$\mathfrak{R}_i = \frac{\eta p A}{H} \tag{3.20}$$

and if we put $H = c'Ad$, where c' is the volume specific heat and d is the thickness of the pyroelectric chip, we get:

$$\mathfrak{R}_i = \frac{\eta p}{c'd}. \tag{3.21}$$

This expression for $\Re_i$ is independent of frequency, as can be seen to the right-hand side of Figure 3.11. It also shows that the current responsivity of a typical detector is inversely proportional to the thickness of the pyroelectric chip.

For frequencies well below the thermal time constant, equation 3.17 may be reduced to:

$$\Re_i = \frac{\eta p A \omega}{G_T}, \tag{3.22}$$

showing that the current responsivity is proportional to frequency and is independent of the thickness of the pyroelectric chip. The fact that it is proportional to frequency means that there is no pyroelectric signal when the incident radiation is not modulated.

Again, if G_T is proportional to the detector area, we can write equation 3.22 in terms of the thermal conductance, g, per unit area:

$$\Re_i = \frac{\eta p A \omega}{g A}, \tag{3.23}$$

which reduces to:

$$\Re_i = \frac{\eta p \omega}{g}, \tag{3.24}$$

and is independent of both the thickness and the area of the pyroelectric chip.

3.6.2. Voltage Responsivity

As explained in Section 3.3, it is normal practice to connect the pyroelectric detector to a high input impedance amplifier. In this case the voltage across the detector is measured, rather than the current generated. An equivalent circuit of the detector illustrated in Figure 3.5 can be drawn as shown in Figure 3.12.

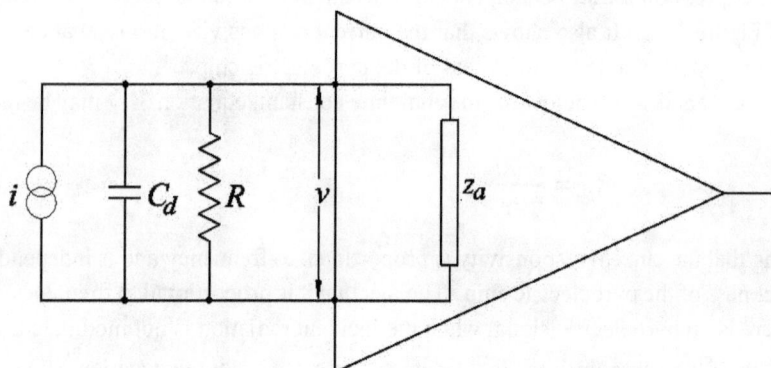

Figure 3.12 Equivalent circuit for detector and amplifier input.

The pyroelectric current, i, produces a voltage across the impedance, Z, of the parallel combination of the detector capacitance, C_d, the bias resistor, R, and the input impedance, Z_a, of the amplifier[2]; i.e.:

$$v = iZ \tag{3.25}$$

Now

$$\frac{1}{Z} = \frac{1}{R} + j\omega C_d + \frac{1}{Z_a} \tag{3.26}$$

The amplifier input impedance, Z_a, can be represented as a parallel combination of capacitance, C_a, and resistance, R_a. However, R_a will generally be much larger than the bias resistor, R, so we can ignore R_a and make the approximation:

$$Z_a = \frac{1}{j\omega C_a} \tag{3.27}$$

So that equation 3.26 becomes:

---

[2] Later, when considering noise, we will also include an additional term G_E, equal to $\omega C_d \tan\delta$. However, as $\tan\delta$ is always much smaller than 1 for a good pyroelectric material, we can ignore G_E compared with $j\omega C_d$, and we will leave it out at this stage.

$$\frac{1}{Z} = \frac{1}{R} + j\omega C_d + j\omega C_a \tag{3.28}$$

i.e.:

$$\frac{1}{Z} = \frac{1+j\omega R}{R} \tag{3.29}$$

or:

$$\frac{1}{Z} = \frac{1+j\omega\,_E}{R} \tag{3.30}$$

where $C = C_d + C_a$, and $\tau_E = RC$ is the electrical time constant.
The voltage responsivity is defined by:

$$\Re_v = \left|\frac{v}{W}\right| = \left|\frac{iZ}{W}\right| \tag{3.31}$$

Combining this with 3.14 gives:

$$\Re_v = \Re_i|Z|. \tag{3.32}$$

from 3.30 we get

$$\left|\frac{1}{Z}\right| = \frac{\left(1+\omega^2\tau_E^2\right)^{1/2}}{R} \tag{3.33}$$

so

$$|Z| = \frac{R}{\left(1+\omega^2\tau_E^2\right)^{1/2}} \tag{3.34}$$

Combining 3.17, 3.32, and 3.34 we get:

$$\Re_v = \frac{\eta p A R \omega}{G_T\left(1+\omega^2\tau_T^2\right)^{1/2}\left(1+\omega^2\tau_E^2\right)^{1/2}} \tag{3.35}$$

Plotting $\Re_v$ as a function of ω gives the frequency dependence of the voltage responsivity, which takes the form shown in Figure 3.13, in which it is assumed that $\tau_E \gg \tau_T$.

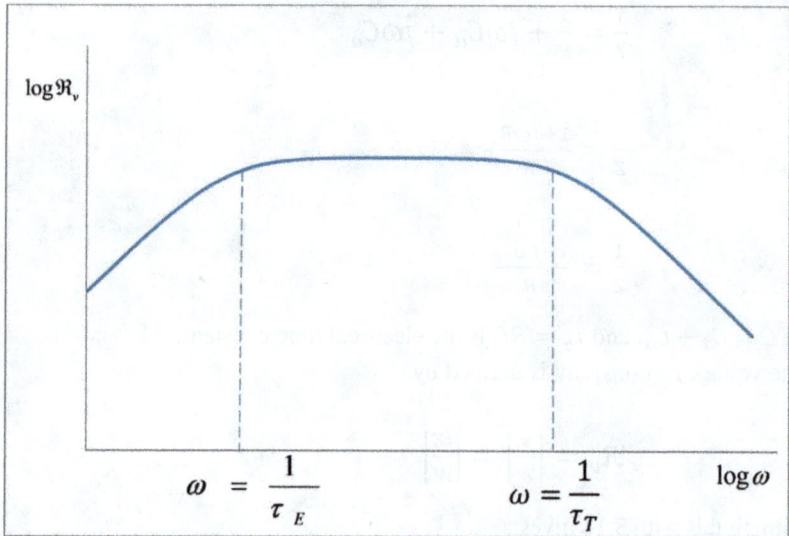

Figure 3.13 Voltage responsivity as a function of frequency.

Let us consider the three regions of this frequency response.

At low frequencies, such that both $\omega\tau_T$ and $\omega\tau_E$ are much less than 1, equation 3.35 reduces to

$$\Re_v = \frac{\eta p A R \omega}{G_T} \tag{3.36}$$

Comparing this with equation 3.22, we see that, as might be expected, $\Re_v$ is equal to $\Re_i R$ and is proportional to frequency.

At high frequencies, such that both $\omega\tau_T$ and $\omega\tau_E$ are much greater than 1, equation 3.35 reduces to

$$\Re_v = \frac{\eta p A R}{G_T \omega \tau_T \tau_E} \tag{3.37}$$

and $\Re_v$ is inversely proportional to frequency.

As before, we can simplify this by putting $\tau_T = H/G_T$ and $\tau_E = RC$, giving:

$$\mathfrak{R}_v = \frac{\eta p A}{H C \omega} \tag{3.38}$$

and, if $H = c'dA$, and $C_d >> C_a$ so that $C = C_d = \varepsilon_0 \varepsilon_r A/d$, then

$$\mathfrak{R}_v = \frac{\eta p}{c' \varepsilon_0 \varepsilon_r A \omega} \tag{3.39}$$

where ε_0 is the permittivity of free space and ε_r is the relative permittivity of the pyroelectric material. So, in this case, the responsivity is independent of the thickness of the pyroelectric chip, but inversely proportional to its area.

At intermediate frequencies, there may or may not be a region in which the voltage responsivity varies only weakly with frequency. Typically, the thermal time constant, τ_T, is within the range 0.01 seconds to 10 seconds. The electrical time constant, τ_E, however, can be anywhere between 10^{-12} seconds and 100 seconds, depending on the magnitudes of the detector capacitance and the bias resistor. If a low value resistor is used, then there is a frequency range over which $\varpi\tau_E << 1$ and $\omega\tau_T >> 1$, and equation 3.35 reduces to

$$\mathfrak{R}_v = \frac{\eta p A R}{G_T \tau_T} \tag{3.40}$$

which is equation 3.19 multiplied by R. In other words, the voltage is equal to the current times the resistance, as would be expected, and we have a frequency independent part of the responsivity curve.

In practice, however, low value resistors would only be used for specialised applications such as laser monitoring (see Chapter 6), and for most applications the resistor is chosen to be as large as possible in order to give maximum signal at low frequencies. The electrical time constant is then in the range 10 s to 100 s. This means that the electrical and thermal time constants are relatively close together and the responsivity looks more like Figure 3.14, and there is no significant frequency independent part of the curve. In Figure 3.14 we have labelled the horizontal axis in Hz to give an idea of the frequency response of a typical detector.

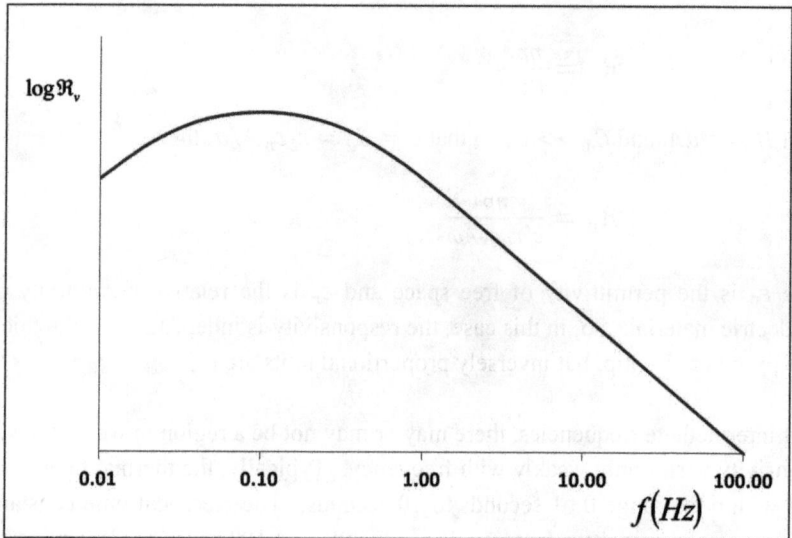

Figure 3.14 Form of voltage responsivity for a typical detector.

3.7. Noise

The usefulness of a detector is usually determined by the smallest incident power that can be detected (minimum detectable power). This is a function of both the responsivity and the noise generated in the detector and the amplifier to which it is connected. An analysis of noise sources is therefore necessary for a full understanding of pyroelectric detectors.

There are, in general, five major noise sources associated with a detector operating with an amplifier in voltage mode. These are represented by the various voltage and current generators in the equivalent circuit shown in Figure 3.15, and will be discussed individually below. Each of the noise sources is defined in terms of root mean square (RMS) current or voltage in unit bandwidth, known as the noise spectral density. In order to calculate the total noise at the input to the following amplifier we will convert the noise voltages into equivalent noise currents. The total mean square noise current can then be obtained from the sum of the squares of the individual noise currents. This total noise current can then be converted back to an effective noise voltage across the parallel combination of detector impedance and amplifier input impedance, as explained for voltage responsivity in section 3.6.2.

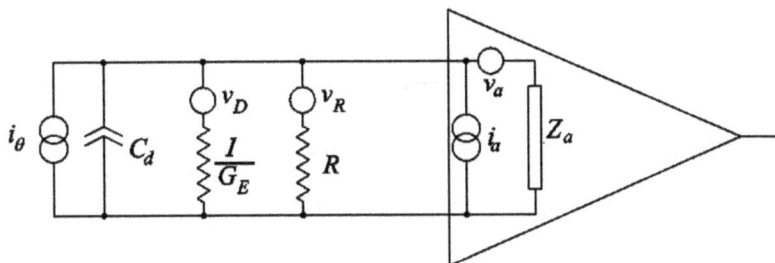

Figure 3.15 Schematic diagram showing noise sources in detector and amplifier input.

3.7.1. Thermal Noise

If a small object of thermal capacity H is connected to a large heat sink at a temperature T by a thermal conductance G_T the mean power flow between the body and the heat sink will be zero when thermal equilibrium has been established, but there will be a randomly fluctuating power flow with an RMS value given by the expression [19] [5]:

$$\Delta W_t = (4kT^2 G_T)^{1/2} \tag{3.41}$$

To obtain an expression for the RMS thermal noise current we can put $W_0 = \Delta W_t$ into the modulus of equation 3.13; but in this case, the power fluctuation is not due to radiant energy, so we put $\eta = 1$. This gives us the following expression for i_T, the thermal noise current:

$$i_T = \frac{pA(4kT^2 G_T)^{1/2} \omega}{(G_T^2 + \omega^2 H^2)^{1/2}} \tag{3.42}$$

Rearranging this we get:

$$i_T = \frac{\eta p A \omega}{\left(G_T^2 + \omega^2 H^2\right)^{1/2}} \frac{\left(4kT^2 G_T\right)^{1/2}}{\eta} \tag{3.43}$$

which is, from equation 3.16:

$$i_T = \Re_i \frac{\left(4kT^2 G_T\right)^{1/2}}{\eta} \tag{3.44}$$

3.7.2. Resistor Noise

The bias resistor, R, will have an associated Johnson noise voltage given by [20] [17]:

$$v_R = (4kTR)^{1/2} \tag{3.45}$$

Applying Ohm's law, this gives an equivalent noise current through the resistor of:

$$i_R = \left(\frac{4kT}{R}\right)^{1/2} \tag{3.46}$$

3.7.3. Dielectric Noise

The pyroelectric detector chip may be considered as a capacitance, C_d, with an associated dielectric loss tangent, $\tan \delta$ [17]. This dielectric loss is represented in Figure 3.15 by an equivalent electrical conductance, G_E, equal to $\omega C_d \tan \delta$.

There will be a Johnson noise voltage associated with this:

$$v_D = \left(\frac{4kT}{\omega C_d \tan \delta}\right)^{1/2} \tag{3.47}$$

equivalent to a noise current:

$$i_D = (4kT \omega C_d \tan \delta)^{1/2} \tag{3.48}$$

3.7.4. Amplifier Noise

The noise at the input of an amplifier may be represented by two noise generators: i_a, the equivalent input current noise, and v_a, the equivalent input voltage noise. Values for these two parameters are usually given by the amplifier manufacturer.

3.7.5. Total Noise

The total mean square noise voltage at the input of the amplifier is now given by:

$$v_n^2 = v_T^2 + v_r^2 + v_d^2 + v_i^2 + v_a^2 \qquad (3.49)$$

where $v_T = i_T|Z|$, $v_r = i_R|Z|$, $v_d = i_D|Z|$, $v_i = i_a|Z|$, and $|Z|$ is defined by equation 3.34. Thus we have, from equations 3.32, 3.34, 3.44, 3.46, 3.48, and 3.49:

$$v_n^2 = \frac{\Re_v^2 4kT^2 G_T}{\eta^2} + \frac{4kTR}{(1+\omega^2\tau_E^2)} + \frac{4kT\omega R^2 C_d \tan}{(1+\omega^2\tau_E^2)} + \frac{i_a^2 R^2}{(1+\omega^2\tau_E^2)} + v_a^2 \qquad (3.50)$$

This expression gives the mean square noise voltage in unit bandwidth at angular frequency ω. This is in units of V^2 Hz^{-1}, so v_n will be in units of V $Hz^{-1/2}$. The amplifier noise voltage, v_a, appears directly in equation 3.50 because it represents noise sources that are independent of the circuit connected to the input of the amplifier, and so is not modified by Z.

An indication of the relative magnitudes of the different noise sources is given in Figure 3.16. This is typical data for a 1 mm square pyroelectric ceramic detector chip connected to a custom-designed CMOS amplifier. As can be seen from the figure, Johnson noise, v_r, in the bias resistor is the dominant noise source at low frequencies, dielectric noise, v_d, is dominant around 10 Hz, and amplifier voltage noise, v_a, is dominant at high frequencies.

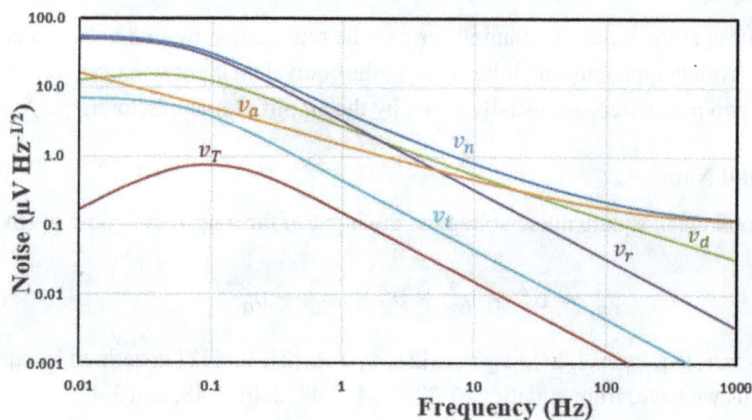

Figure 3.16 Noise voltages in a typical pyroelectric detector.

The thermal time constant of this detector is approximately 1 second and the electrical time constant is about 2.5 seconds.

3.7.6. Non-Gaussian Noise

To complete the picture, we must also consider certain "non-Gaussian" noise sources that are not so easily modelled as the five noise sources described above.

Thermally induced noise (TIN) and thermally induced transients (TITs) can occur when the ambient temperature of a detector changes. For example, the sudden onset of TITs has been observed in lithium tantalate detectors when the temperature of the detector is slowly changed. Initially, when the temperature is slowly raised by one or two degrees, the noise is Gaussian and complies with the analysis above, but as the temperature continues to rise, after two or three degrees increase there is a sudden onset of large, random, but frequent noise spikes. If the temperature change is then reversed there is again a period of low noise followed by a sudden onset of large noise spikes. Possible explanations for this effect include domain realignment within the material. Polycrystalline ceramics are generally less prone to these effects than single crystals.

Shot noise, when there is an isolated single spike of noise, may be induced by alpha particles or high energy cosmic rays.

Random telegraph signals (RTS), which are characterised by randomly occurring spikes, result from instantaneous changes to the FET gate leakage current over time scales ranging from a few milliseconds to weeks. This is thought to originate either from single atom impurities (e.g. gold) adjacent to the gate structure, which gain and lose thermal electrons and modulate the leakage current, or from defects in the semiconductor itself [21].

When used in applications such as intruder alarms, where a response to a single event is required following a long period of inactivity, it is important to consider ways to mitigate the effects of non-Gaussian noise. One approach is to arrange the optics and positioning of the detector so that a repeating or alternating signal is produced, as described in section 6.2.1. Another approach is to view the scene with two (or more) detectors and respond only to events that occur in both detectors simultaneously.

3.8.   Figures of Merit

There are several figures of merit which are widely used to quantify the relative performances of pyroelectric detectors (and, indeed, many other types of radiation detector) in terms of their responsivity and noise.

3.8.1. Noise Equivalent Power

The most commonly used figure of merit is the Noise Equivalent Power, or NEP, which is defined as the incident power which is required to produce a signal equal to the RMS noise voltage, i.e.:

$$NEP = \frac{v_n}{\Re_v} \qquad (3.51)$$

As the responsivity is frequency dependent and the noise is bandwidth dependent, the NEP is specified for a particular frequency and bandwidth, and is measured in units of W Hz$^{-1/2}$. Occasionally, however, the broadband NEP is quoted, in Watts; this is the ratio of the broadband noise (over a specified bandwidth) to the responsivity at a particular frequency.

3.8.2. Noise Equivalent Irradiance

It is sometimes more instructive to consider the Noise Equivalent Irradiance (NEI) of a detector, which is a measure of its ability to detect a given incident power density, or irradiance. The NEI is defined by:

$$NEI = \frac{v_n}{\Re_v A} \tag{3.52}$$

and is measured in units of W m^{-2} Hz$^{-1/2}$.

3.8.3. Detectivity

The NEP of a detector, being an inverse signal to noise ratio, gets smaller as the detector gets better. However, it is often considered to be aesthetically more pleasing to have a figure of merit which is larger for better detectors. For this reason, the term Detectivity, D, has been introduced, which is the reciprocal of the NEP.

A more commonly used term, however, is the specific detectivity, D^*, which is defined as

$$D^* = \frac{A^{1/2}}{NEP} \tag{3.53}$$

The reason for defining D^* in terms of the square root of the detector area is that it was introduced as a measure of the performance of photodetectors in which the NEP is largely proportional to $A^{1/2}$. In this case, D^* is independent of the detector area and is a good measure of the comparative performance of detectors of different areas. In the case of pyroelectric detectors, however, D^* is only independent of area if dielectric noise is dominant, which is true in only a very limited number of cases.

Historically, D^* is expressed in *Jones units,* i.e. cm Hz$^{1/2}$ W^{-1} in honour of R. Clark Jones [22] [23] who originally defined it, although m Hz$^{1/2}$ W^{-1} would be more correct in the SI system of units.

3.9. Performance Examples

To give an idea of the order of magnitude of the various performance parameters, the following figures illustrate the typical values for a 1 mm square 200 µm thick pyroelectric ceramic detector connected to a custom designed CMOS amplifier. A germanium window with a 7.5 µm long wave pass filter coating has been assumed, giving a total transmission of 48% for 20 °C blackbody radiation. More specific examples will be given in later chapters.

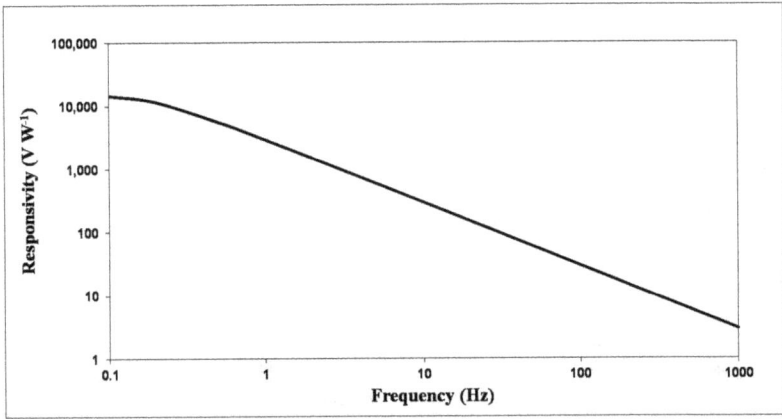

Figure 3.17 Responsivity of 1 mm square pyroelectric detector

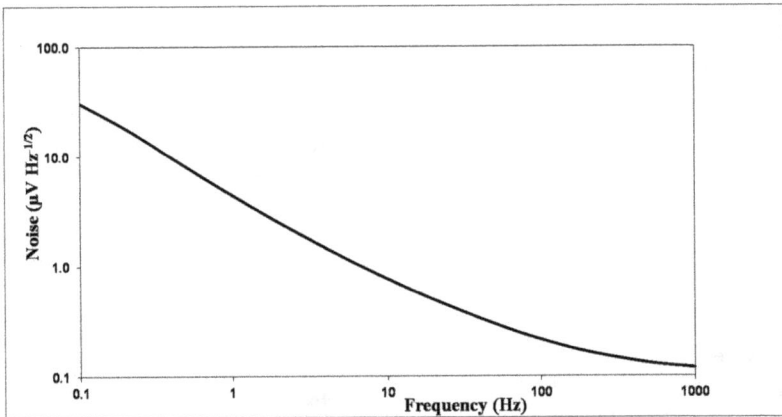

Figure 3.18 Noise of 1 mm square pyroelectric detector

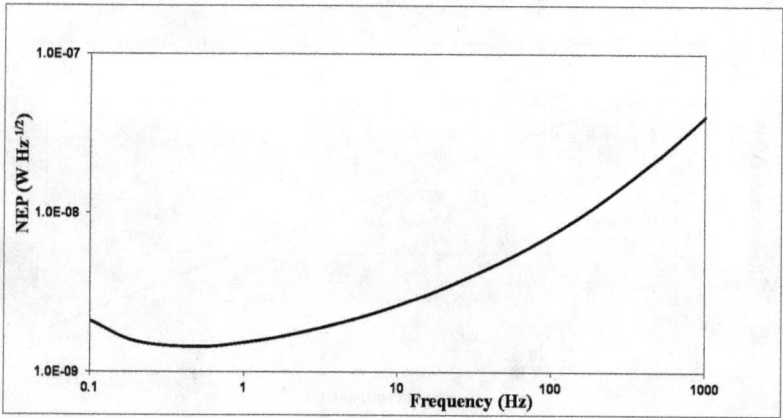

Figure 3.19 Noise Equivalent Power of 1 mm square pyroelectric detector

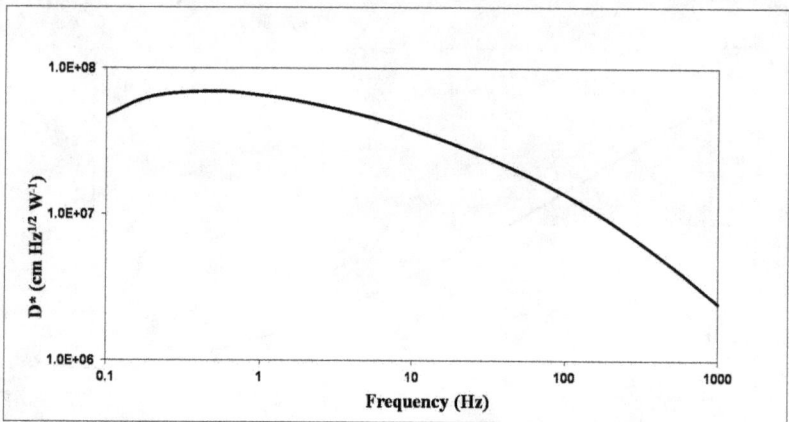

Figure 3.20 Specific Detectivity of 1 mm square pyroelectric detector

3.10. Microphony

One other important factor affecting the performance of pyroelectric detectors is microphony. All pyroelectric detectors are piezoelectric, which means that a mechanical strain in the material will result in an electrical charge being generated. This means that any vibration experienced by a pyroelectric detector in its eventual application is likely to generate an additional unwanted signal or noise. For most pyroelectric materials, there are two piezoelectric coefficients which contribute most of the piezoelectric response. These are d_{33} and d_{31}; d_{33} gives a response due to the self-loading of the element through its thickness under the effects of vibrations perpendicular to the element plane. Generally, this is rather weak. d_{31} gives a response due to stretching of the detector element in its plane, which can be caused either by in-plane accelerations, or by flexure of the substrate on which it is mounted. The magnitude of this can be very significant, and if vibration of the detector is likely to be an issue, then serious consideration must be given to the mechanical design of the detector mount, to minimise substrate flexure under device vibration and the ways any flexure might couple to the pyroelectric element.

4. Pyroelectric Materials

As mentioned at the beginning of Chapter 2, the naturally occurring pyroelectric material tourmaline has been known for hundreds of years [1], and the phenomenon of pyroelectricity was studied in materials such as Rochelle salt in the nineteenth century [3]. The known ferroelectric materials now amount to several hundred [24], some of which have been used in a variety of applications including piezoelectric transducers for loudspeakers and microphones, electrooptic devices for optical modulators, and high permittivity materials for capacitors.

For a pyroelectric detector, the primary requirement is a high pyroelectric coefficient and low dielectric loss, usually a low dielectric constant, and a reasonably wide temperature range of operation. The five materials most commonly used in pyroelectric detectors are triglycine sulphate (TGS) [25], lithium tantalate (LT) [26] [27], strontium barium niobate (SBN) [28], polyvinylidene fluoride (PVDF) [29] [30], and ceramic materials based on lead zirconate titanate (PZT) [31] [18]. Typical values, at 20°C unless otherwise stated, of some of the relevant properties of these materials are listed in Table 4.1. We will briefly discuss each of these materials in turn.

Material	TGS	LT	SBN	PVDF	PZT
Pyroelectric coefficient, p (10^{-4} C m^{-2} K^{-1})	2.8	1.8	6.5	0.3	3.5
Relative permittivity, ε_r	38	54	380	10	250
Dielectric loss, $tan\,\delta$	0.01	0.003	0.003	0.03	0.005
Volume specific heat, s (10^6 J m^{-3} K^{-1})	2.3	3.3	2.3	2.4	2.6
Curie temperature, T_c (°C)	49	620	116	>100	200
Resistivity, ρ (Ω m)	10^{13}	10^{13}	10^{10}	10^{14}	10^{10}

Table 4.1 Properties of some pyroelectric materials

4.1. Triglycine Sulphate

Triglycine sulphate (TGS) is a colourless, water-soluble crystal that is grown from solution; its chemical formula is $(NH_2CH_2COOH)_3H_2SO_4$. It is hygroscopic and rather fragile, so that special precautions are necessary when it is being processed, and its low Curie temperature is a major disadvantage. Despite these problems, however, TGS has been used commercially in applications where sensitivity is of prime importance; it offers the highest detectivity for all but very small detectors operating at low frequencies.

The low Curie temperature of 49 °C is a serious disadvantage of TGS. Not only does the pyroelectric effect disappear if the temperature rises above 49 °C, but the material must be re-poled by the application of an electric field once the temperature decreases again. Two possible solutions to this problem have been investigated. The first is to replace the hydrogen atoms in TGS with deuterium, which raises the Curie Temperature to about 60 °C [32]. This, however, results in a decreased pyroelectric coefficient at 20 °C.

The second approach is to dope TGS with L-alanine [33]. As illustrated in Figure 4.1, the molecular structure of L-alanine is more asymmetric than glycine, which means that the crystal retains its polarization after a temperature excursion above the Curie temperature.

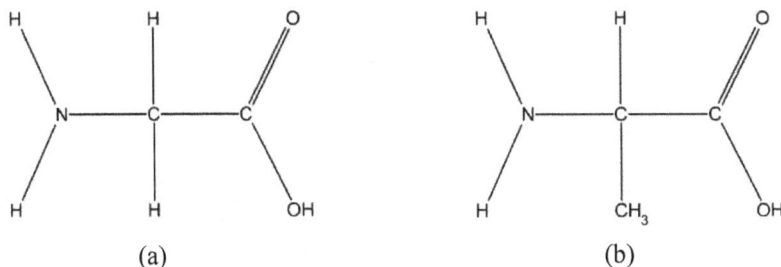

Figure 4.1 Molecular structure of (a) glycine and (b) L-alanine

However, despite the availability of these modifications, TGS has very limited applications because of its fragile, hygroscopic nature.

4.2. Lithium Tantalate

Lithium tantalate, $LiTaO_3$, in contrast with TGS. is an insoluble material with a very high Curie temperature. This means that its performance as a pyroelectric detector, though inferior to TGS, is independent of temperature over a wide range. It can also be handled in thinner slices than TGS, thus regaining some performance. Good single crystals of lithium tantalate can be grown by the Czochralski technique. Lithium tantalate is widely used, in single and dual element pyroelectric detectors, in intruder alarms.

4.3. Strontium Barium Niobate

Strontium barium niobate (SBN) is a family name for a range of solid solutions defined by the formula $Sr_{(1-x)}Ba_xNb_2O_6$ in which x can be varied within the range 0.25 to 0.82. The properties listed in Table 4.1 are for x = 0.52. In general, the Curie temperature increases and the pyroelectric coefficient at room temperature decreases as x is increased. Thus, by varying the composition of SBN it is possible to tailor the properties to particular applications.

Like lithium tantalate, SBN crystals are produced by the Czochralski technique, but good quality crystals large enough for pyroelectric detector production are difficult to grow, so SBN is not widely used in this application.

4.4. Polyvinylidene Fluoride

Polyvinylidene fluoride (PVDF) is a plastic film similar to polyethylene. The pyroelectric phase is produced by stretching the film and applying an electric field. It is not a true ferroelectric and does not have a well-defined Curie temperature. Although PVDF has a relatively low pyroelectric coefficient, it also has a low relative permittivity and is useful for large area pyroelectric detectors. It is sometimes used for laser pulse energy monitors, where a large area detector is required and the high power output from a laser means that low responsivity is not a problem.

The performance of PVDF as a pyroelectric detector is generally inferior to the other materials mentioned here, except for very large area devices, and microphony can be a problem. It does, however, have the advantage of very low cost and is available in large thin sheets, which do not require any slicing, lapping, or polishing.

4.5. PZT Pyroelectric Ceramics

Although single crystals, oriented with the polar axis perpendicular to applied electrodes, offer the best pyroelectric performance, certain materials can be used as polycrystalline ceramics. This is particularly true of materials, like barium titanate, in which the paraelectric phase (above the Curie temperature) is cubic, and the ferroelectric phase is tetragonal, orthorhombic, or rhombohedral. In the tetragonal case, for example, there are six possible polar directions so that, if an electric field is applied in a random direction, it cannot be more than 55° from one of the possible polar directions in any of the ceramic grains. This means that 83 % of the single crystal polarisation is theoretically obtainable in a ceramic of randomly oriented crystallites. For rhombohedral and orthorhombic structures there are 8 and 12 available polar directions, and an even higher percentage of the single crystal polarisation may be obtained.

The major advantage of a ceramic is that it is relatively easy to produce large blocks of uniform material. This is generally achieved by sintering the mixed oxide powders at temperatures of the order of 1000 °C. For the best quality ceramics, however, hot-pressing techniques are employed [34] [18].

These blocks of ceramic can be cut, lapped, polished, and diced by processes very similar to those employed in the manufacture of semiconductor devices. There is no need to consider orientation during this processing because the ceramic can be poled in any direction by the application of a suitable electric field at an elevated temperature.

There is a vast range of ceramic compositions consisting of solid solutions of lead zirconate, lead titanite, and a variety of other oxides. These have been developed over many years for various applications utilizing their ferroelectric, piezoelectric, electro-optic, and pyroelectric properties. The figures listed under PZT in Table 4.1 are for a typical composition with a high ratio of lead zirconate to lead titanate. Compositions including small amounts of iron and niobium have been developed specifically for pyroelectric detectors [35] [36], for which the major requirements are a high pyroelectric coefficient and a low relative permittivity.

Another advantage of PZT ceramics is that their electrical resistivity can be controlled within the range 10^9 Ωm to 10^{11} Ωm by the addition of small amounts of uranium [18]. This means that the FET bias resistors may be omitted from detectors, a very significant advantage in the manufacture of arrays of small elements. The conductivity of the ceramic material increases with increasing temperature, compensating nicely for the increasing leakage current of the FET.

4.6. Comparative Performance

The choice of material for a pyroelectric detector is not always a straightforward matter; it depends on the size of the required detector element and the intended frequency of operation. Environmental conditions (particularly temperature range), maximum incident power, and cost of manufacture must also be considered.

In order to obtain an indication of the relative merits of different materials, specific detectivities have been calculated using equations 3.35, 3.50, 3.51, and 3.53. with the values listed in Table 4.1. This has been done for the frequency range 0.1 Hz to 1 kHz and for three detector element areas: 100 mm^2, 1 mm^2, and 0.01 mm^2. A detector thickness of 50 μm has been assumed, and the values of η and R have been taken as 1.0 and 5 x 10^{11} Ω respectively. The temperature is taken as 290 K.

The thermal conductance, G_T, is assumed to be proportional to the detector element area, i.e. $G_T = gA$, and g has been assigned the value 1.8 x 10^3 W K^{-1} m^{-2}. This value accords with that which is found in typical detectors mounted on alumina substrates.

For the amplifier input we have assumed a MOSFET with an input capacitance of 2 pF, a input current noise of 4.4 x 10^{-17} A Hz$^{-1/2}$, and a frequency dependent voltage noise given by the formular:

$$v_a^2 = 1.5 \times 10^{-14} + \frac{2.5 \times 10^{-12}}{\omega}$$

                                                                    4.1

The results are shown in Figures 4.2 to 4.4.

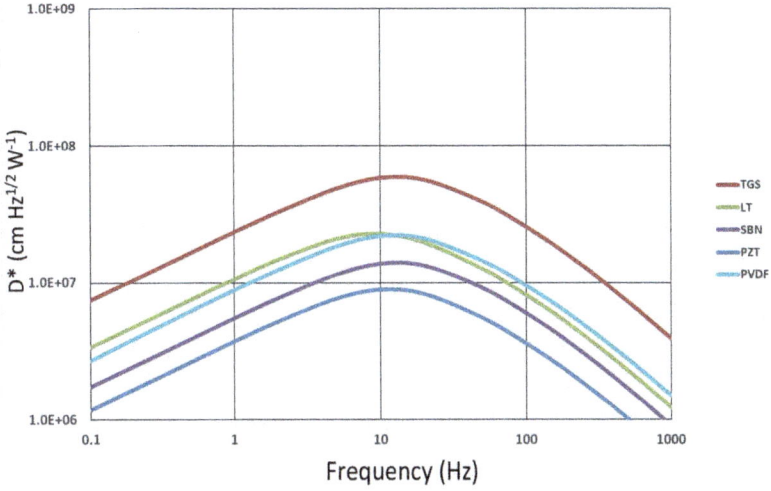

Figure 4.2 Detectivity with different materials for 100 mm² detector

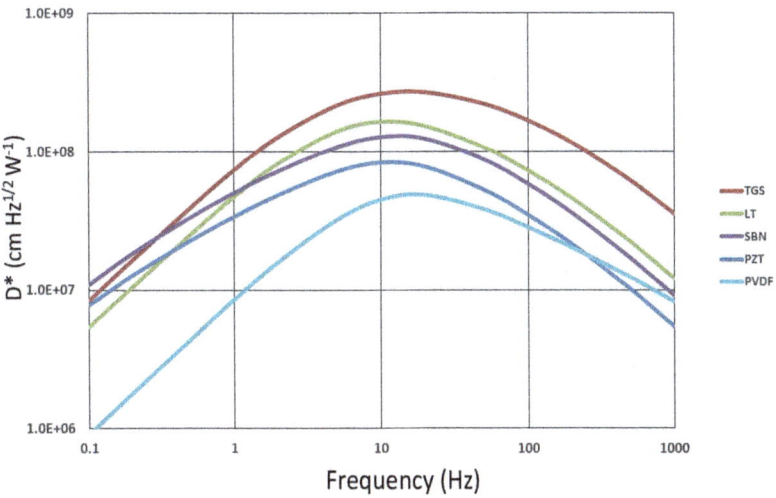

Figure 4.3 Detectivity with different materials for 1 mm² detector

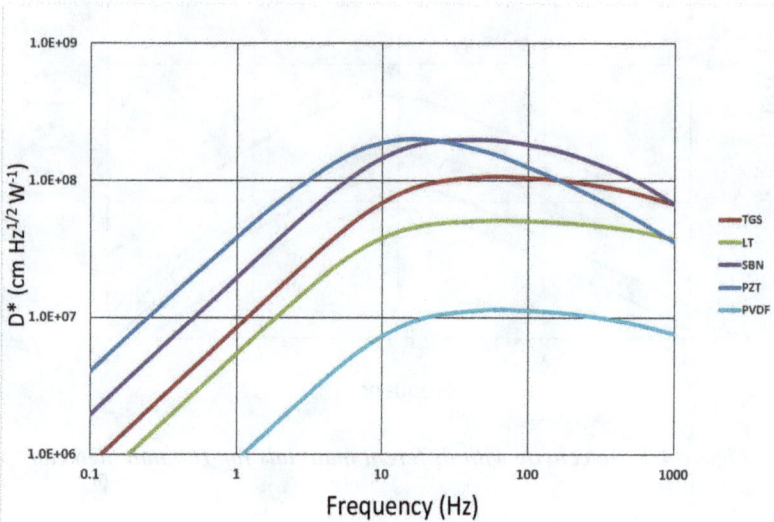

Figure 4.4 Detectivity with different materials for 0.01 mm² detector

As may be seen from these three graphs, the performance of TGS is significantly better than any other material for a large detector of 100 mm² or above, with PVDF and LT being approximately equal but a poor second. This is representative of the size of detector used in passive infrared intruder alarms, but because of the other disadvantages of TGS, the majority of intruder alarms use either lithium tantalate or a variant of PZT.

At the other extreme, for a 0.01 mm² detector, as might be used in an array, PZT is significantly better than both TGS and LT for frequencies up to at least 100 Hz, and PVDF is very inferior to all the other materials. Arrays of small elements therefore tend to use PZT. SBN is not used because of the difficulty of manufacture.

5. Infrared Detection Devices

5.1. Introduction

In Chapter 3 we discussed the basic concepts of pyroelectric detectors and introduced a typical single element device. In this chapter we will examine devices in more detail, starting with the single element detector and progressing to multi-element two dimensional arrays.

5.2. Single Element Detectors

In Chapter 3 we described a single element pyroelectric detector and illustrated it in Figures 3.4 and 3.8, which are reproduced here as Figure 5.1 and Figure 5.2. This basic format is widely used, although different manufacturers have employed different variations on this theme.

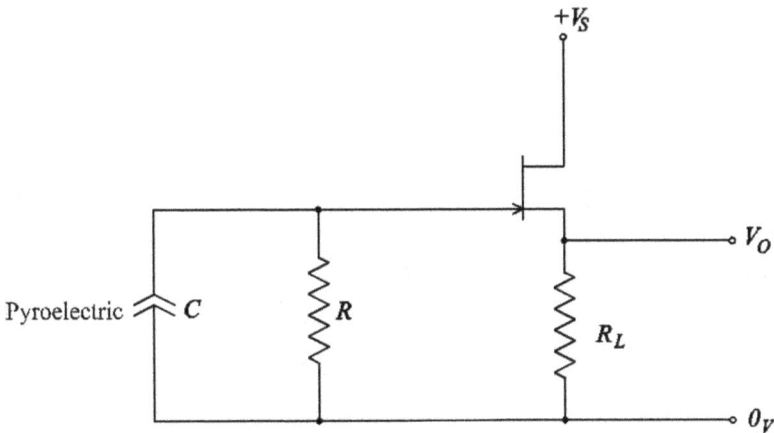

Figure 5.1 Schematic diagram of a pyroelectric detector with a JFET.

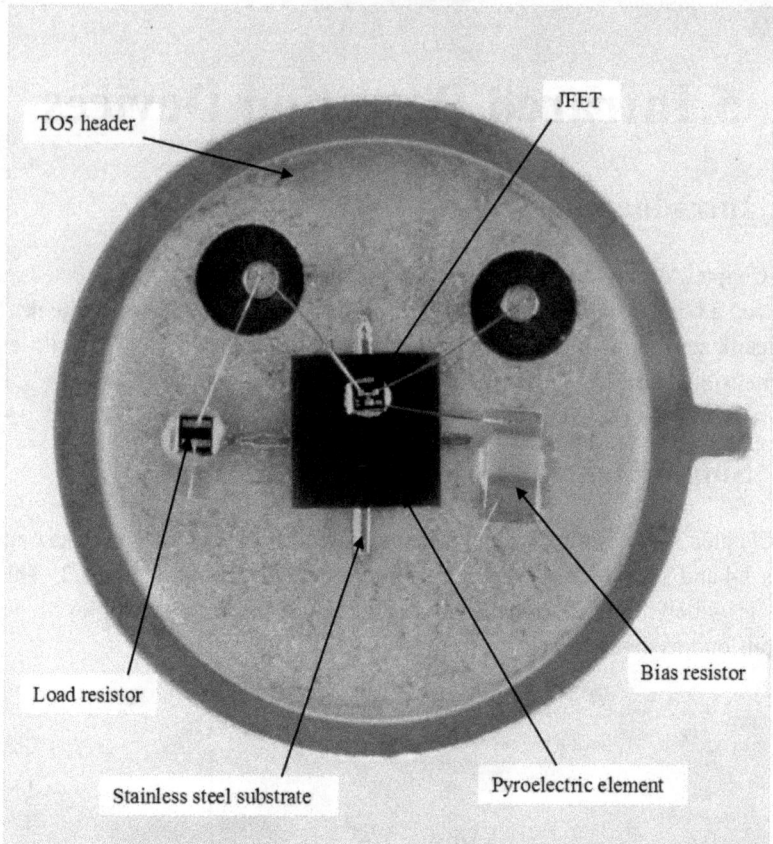

Figure 5.2 Photograph of the internal structure of a pyroelectric detector.

5.2.1. Alternative Bias Arrangements

5.2.1.1. Diode Biasing

The aim of an infrared detector is to detect small variations in the input irradiance, resulting in temperature changes of millidegrees, or even micro-degrees in the pyroelectric element. Consequently, if a large input signal, or a change in the ambient temperature of a degree or more, is experienced, the detector will be overloaded, and may take some time to recover. In order to overcome this problem, some manufacturers [37] have replaced the high value bias resistor with one or more non-linear components.

One example of this is illustrated in Figure 5.3, in which the bias resistor is replaced by a pair of low leakage diodes.

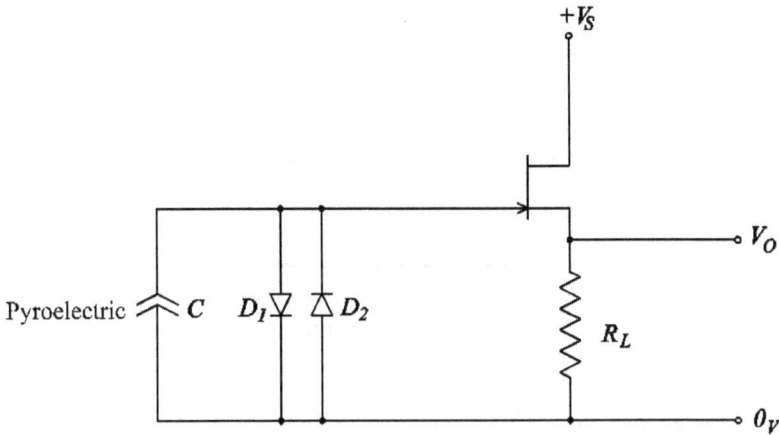

Figure 5.3 Pyroelectric detector with bias resistor replaced by two diodes.

The relationship between the voltage applied across a diode and the current flowing through it is given by an expression of the form:

$$I = I_0\left(exp^{eV/kT} - 1\right) \tag{5.1}$$

where I is the current flowing through the diode when the voltage applied is V, I_0 is the leakage current under reverse bias, e is the charge of an electron, k is Boltzmann's constant, and T is the absolute temperature. A low leakage diode will have a reverse bias leakage current of the order of 10 pA, and the current versus voltage $(I - V)$ characteristic of such a device at a temperature of 300 K will be of the form shown in Figure 5.4.

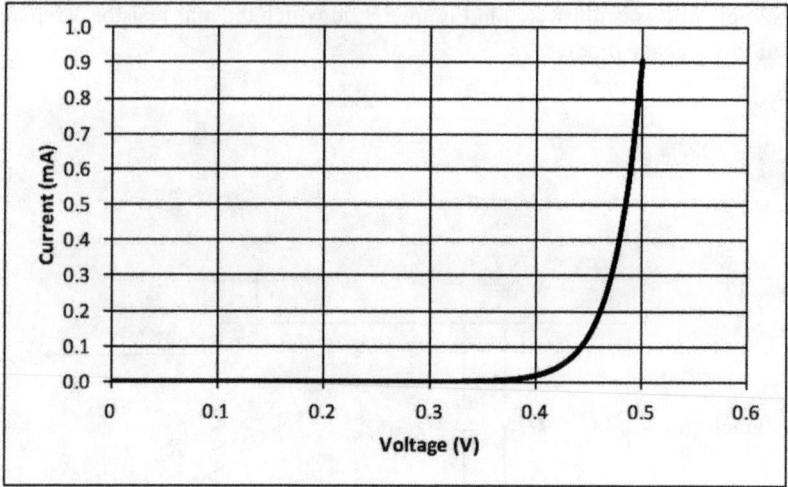

Figure 5.4   I-V characteristic of a low leakage diode.

In the case of a parallel pair of diodes, as in Figure 5.3, the $I - V$ characteristic becomes symmetrical, as shown in Figure 5.5.

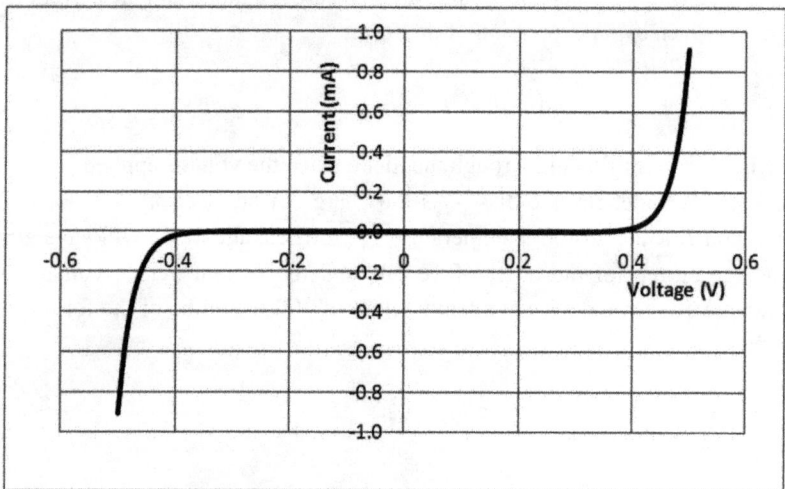

Figure 5.5   I-V characteristic of a parallel pair of low leakage diodes.

Thus, the effective resistance of this pair of diodes is greater than 10^9 Ω for bias voltages of less than ± 80 mV, but reduces to 10^6 Ω for a bias voltage of about ± 300 mV, and to 10^3 Ω at about ± 500 mV. So, for small signals of the order of a few millivolts, the detector will operate in similar manner to a detector with an ohmic bias resistor, but large signals will be strongly attenuated.

Another advantage of this arrangement is that a pair of diodes is generally of lower cost than a high value bias resistor. It also offers the possibility of integrating the JFET, the two diodes and the load resistor into a single semiconductor chip, thus reducing the component count and further reducing the cost of the assembly.

Although this arrangement works well at preventing amplifier overload, it results in a detector with a very non-linear voltage responsivity. Also, if the ambient temperature is changing, so that the resultant pyroelectric current causes the bias voltage to deviate from zero, the responsivity will decrease due to the lower effective resistance of the diode pair. This behaviour may be acceptable if the detector is used purely for detection purposes (although even then the variable sensitivity may be a problem) but it would not be acceptable if any form of quantitative assessment of the infrared signal is required, as would be the case if the detector were being used for remote temperature measurement.

5.2.1.2. Electrically Conducting Pyroelectric Material

In Chapter 4 we have described how some manufacturers have developed pyroelectric ceramic materials that have an electrical resistivity within the range 10^9 to 10^{11} Ωm. This allows the pyroelectric chip to have a resistance of the order of 10^{12} Ω, and so provide a path for the bias current of the JFET without the need for a separate bias resistor. This is particularly useful in the case of one- or two-dimensional arrays with hundreds, or even thousands of detector elements (see section 5.5), in which it would be very difficult or impossible to incorporate separate bias resistors for each element.

A further advantage of the electrically conducting pyroelectric ceramic is that, because it is a semi-insulator, its resistivity decreases with increasing temperature, unlike a normal resistor for which the resistivity increases with increasing temperature. This can compensate, at least to some extent, for the increase in JFET leakage current with temperature.

5.2.1.3. Reset Switching

For some applications it may be possible to leave out the bias resistor altogether and apply a regular reset to the amplifier input. Figure 5.6 illustrates one method of doing this. The bias resistor R in Figure 5.1 is replaced by a second transistor, T_2, which is normally turned off, but is periodically turned on by applying a reset pulse at the point labelled "Reset". The input gate of transistor T_1 will drift over time due to both the leakage current of T_1 and ambient temperature changes producing pyroelectric charges on the detector. A periodic reset pulse will bring the bias voltage back to zero. This technique is particularly applicable when the detector is used to monitor a periodic signal, as produced by a mechanical chopper, in which case the reset pulse can be applied at an appropriate point in the signal: for example, when the signal would be expected to be zero in the absence of any drift. This technique will be discussed in more detail in Chapter 6.

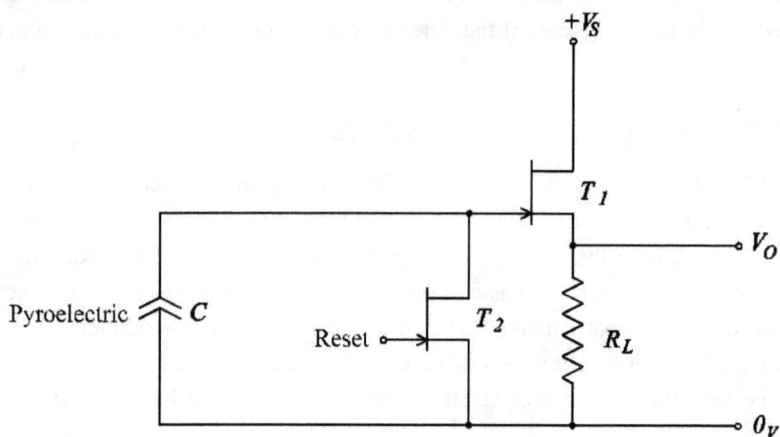

Figure 5.6 Schematic arrangement for applying a periodic reset to a pyroelectric detector.

5.2.2. Edge electrodes

So far, we have confined ourselves to the discussion of single element detectors in which the electrodes are deposited on the large faces of the detector chip, one of which is arranged to absorb the incident radiation. This is illustrated in Figure 3.1, which is reproduced here as Figure 5.7.

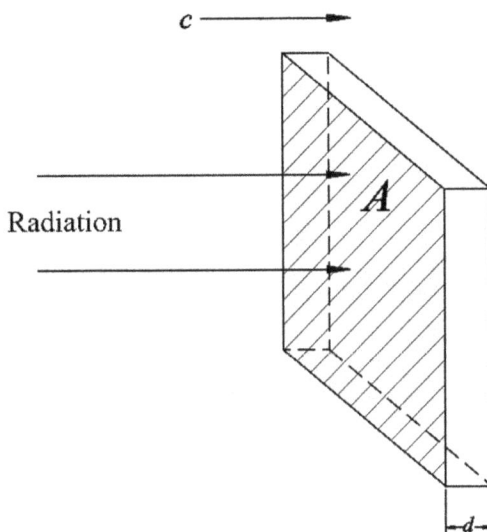

Figure 5.7 Face electrode detector.

An alternative arrangement is to deposit the electrodes on the edges of the pyroelectric chip, as shown schematically in Figure 5.8. The pyroelectric material is now arranged so that the polar axis is perpendicular to the electroded faces.

As mentioned in Section 3.6.2, the electrical capacitance of the detector chip illustrated in Figure 5.7 is given by the expression $C_d = \varepsilon_0 \varepsilon_r A/d$. For the edge electrode detector of Figure 5.8 the capacitance is $C_e = \varepsilon_0 \varepsilon_r A_E/d_E$. If by way of example, the detector chip is 10 mm square and 1mm thick, then C_e will be a factor of 100 smaller than C_d. This can be used in conjunction with a low value resistor to achieve a very short electrical time constant, which makes the detector suitable for detecting very short pulses of radiation, such as from a Q-switched laser. This application will be discussed in more detail in Chapter 6.

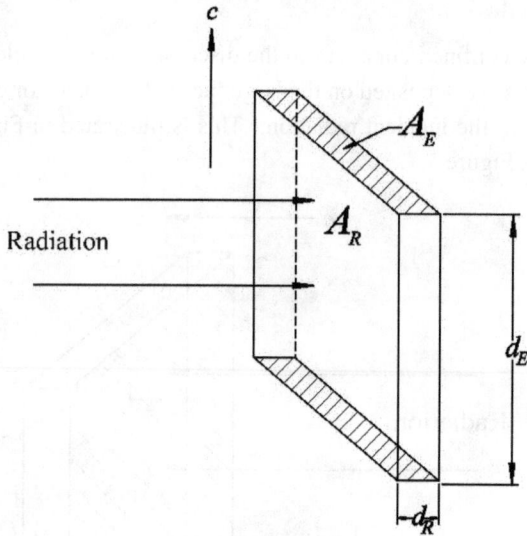

Figure 5.8 Edge electrode detector.

5.3. Compensated Detectors

A pyroelectric infrared detector is designed to detect small changes in incident radiation, which produce temperature changes within the detector material of a few micro-degrees. The pyroelectric detector is therefore inherently very sensitive to small changes in ambient temperature. In fact, if a single element detector is deployed outside without any attempt to shield it from ambient temperature variations, it will be rendered useless by the thermally induced noise.

One widely used solution to this problem is to employ a compensated detector. The basic principle of compensation is to use two similar pyroelectric elements connected in opposite polarity but exposing only one of these two elements to the radiation to be detected at any one time. There are two possible arrangements for a compensated detector, which we will consider separately.

5.3.1. Parallel Compensation

The two pyroelectric elements may be connected in parallel, as illustrated in Figure 5.9.

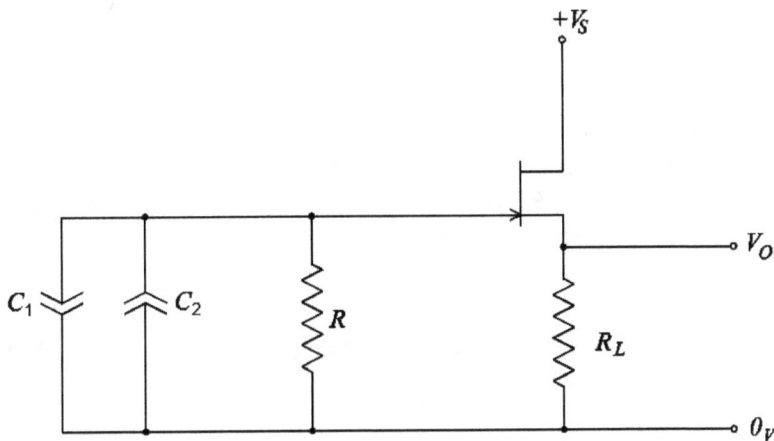

Figure 5.9 Parallel compensation.

For effective compensation, it is required that the same change in temperature occurring in each of the two elements produces zero signal at the input of the amplifier. The pyroelectric currents, i_1 and i_2, generated in the two elements, C_1 and C_2, must be equal and opposite to give zero voltage across the combined capacitance $C_1 + C_2$.

Now equation 3.12 states:

$$i = pA\frac{dT}{dt} \tag{5.2}$$

so we must have:

$$pA_1\frac{dT}{dt} = pA_2\frac{dT}{dt} \tag{5.3}$$

and therefore:

$$A_1 = A_2 \tag{5.4}$$

i.e., the areas, A_1 and A_2, of the two capacitors must be equal.

Although there is no requirement for the two capacitances to be equal, in practice it is likely that the two elements will be parts of the same pyroelectric chip, and therefore be of the same thickness. In this case, if the areas are equal, then the capacitances will be equal, i.e., $C_1 = C_2$.

5.3.1.1. Responsivity and Noise for a Parallel Compensated Detector

To compare the responsivities of compensated and uncompensated detectors, let us start by re-drawing the equivalent circuit of Figure 3.12 using Z_1 to represent the impedance of the detector RC combination, as shown in Figure 5.10. For ease of analysis, we will consider the case of an electrically conducting pyroelectric material so that each element has its own in-built resistor, and we can draw the equivalent circuit of the compensated detector as shown in Figure 5.11. Here Z_1 represents the active element and Z_2 is the compensating element; radiation falls only on the element Z_1, generating a signal current i.

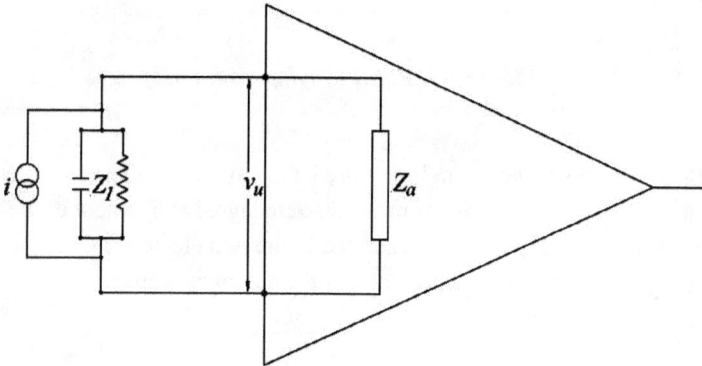

Figure 5.10 Equivalent Circuit for Uncompensated Detector.

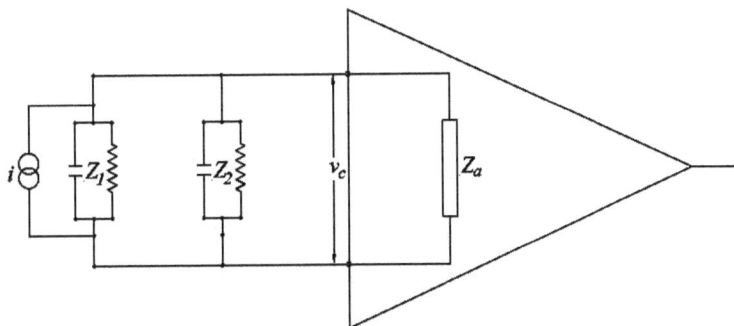

Figure 5.11 Equivalent Circuit for Parallel Compensated Detector.

In the uncompensated case, the voltage seen by the amplifier is simply that generated by the current i flowing through the combined impedance, Z_u, given by:

$$\frac{1}{Z_u} = \frac{1}{Z_1} + \frac{1}{Z_a} \tag{5.5}$$

i.e.

$$Z_u = \frac{Z_1 Z_a}{Z_1 + Z_a} \tag{5.6}$$

so that:

$$v_u = \frac{i Z_1 Z_a}{Z_1 + Z_a} \tag{5.7}$$

In the compensated case, the combined impedance, Z_c, is given by:

$$\frac{1}{Z_c} = \frac{1}{Z_1} + \frac{1}{Z_2} + \frac{1}{Z_a} \tag{5.8}$$

so

$$Z_c = \frac{Z_1 Z_2 Z_a}{Z_1 Z_a + Z_2 Z_a + Z_1 Z_2} \tag{5.9}$$

The current i flows through this combined impedance and generates the voltage

$$v_c = \frac{iZ_1 Z_2 Z_a}{Z_1 Z_a + Z_2 Z_a + Z_1 Z_2} \tag{5.10}$$

We can now combine equations 5.7 and 5.10 to give:

$$\frac{v_c}{v_u} = \frac{Z_2(Z_1 + Z_a)}{Z_1 Z_a + Z_2 Z_a + Z_1 Z_2} \tag{5.11}$$

We have shown above that for effective compensation the active and compensating elements must be equal in area, and therefore of equal impedance if they are formed on the same pyroelectric chip. So, $Z_2 = Z_1$ and equation 5.11 becomes

$$\frac{v_c}{v_u} = \frac{Z_1(Z_1 + Z_a)}{Z_1 Z_a + Z_1 Z_a + Z_1 Z_1} \tag{5.12}$$

i.e.

$$\frac{v_c}{v_u} = \frac{Z_1 + Z_a}{Z_1 + 2Z_a} \tag{5.13}$$

For large area detectors, having a high capacitance, and a high input impedance amplifier, Z_a will be high compared with Z_1 and equation 5.13 approximates to

$$\frac{v_c}{v_u} = \frac{1}{2} \tag{5.14}$$

For small area detectors, however, the input capacitance of the amplifier may be similar to that of the detector element and the impedances are approximately equal. In this case equation 5.13 reduces to

$$\frac{v_c}{v_u} = \frac{2}{3} \tag{5.15}$$

Thus, we see that the responsivity of a parallel compensated detector is at best two thirds and at worst one half of the responsivity of the equivalent uncompensated detector.

As far as noise generation is concerned, a parallel compensated detector having two identical elements connected in parallel can be considered as a single element having twice the capacitance and half the resistance of an uncompensated element, but the same electrical time constant (because $\tau_E = RC$). The total noise is now given by equation 3.50, which we repeat here with R_1 divided by 2 and C_1 multiplied by 2:

$$v_n^2 = \frac{\Re_v^2 4kT^2 G_T}{\eta^2} + \frac{4kT(R_1/2)}{(1+\omega^2\tau_E^2)} + \frac{4kT\omega(R_1/2)^2(2C_1)\tan\delta}{(1+\omega^2\tau_E^2)} + \frac{i_a^2(R_1/2)^2}{(1+\omega^2\tau_E^2)} + v_a^2 \quad (5.16)$$

This tells us that the effect of compensation on the total mean square noise depends on which source of noise is dominant. If thermal noise were dominant, as represented by the first term in equation 5.16, then noise would decrease in line with the responsivity, and the signal to noise ratio would remain the same. The second term, representing resistor noise, is reduced by a factor of 2. The third term, for dielectric noise, is also reduced by a factor 2. The fourth term, amplifier current noise, is reduced by a factor of 4, and the fifth term, amplifier voltage noise, is unchanged. The RMS noise is the square root of this mean square noise, so it is reduced by a factor in the range 1 to 2, depending on which term is dominant.

Thus, we see that, at best signal and noise are reduced by the same amount and the signal-to-noise ration remains the same, and at worst the signal is halved while the noise is unchanged, so the signal to noise ratio is degraded by a factor of two. The precise result, between these two limits, depends on a variety of factors, including the pyroelectric material, the structure and size of the detector, and the frequency range of operation.

5.3.2. Series Compensation

An alternative compensating arrangement is to connect the two elements in series, as illustrated in Figure 5.12. In this case the analysis is a little more complicated, but if we again take the case of an electrically conducting pyroelectric material, each element has its own inbuilt resistor, and we can re-draw the equivalent circuit as shown in Figure 5.13.

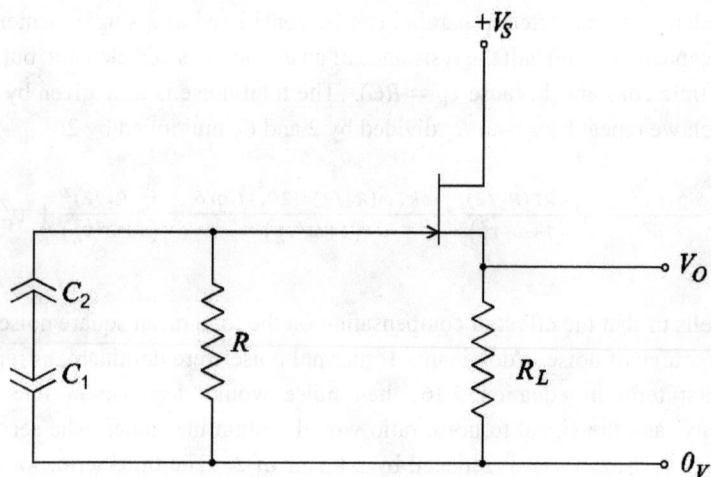

Figure 5.12 Series compensation.

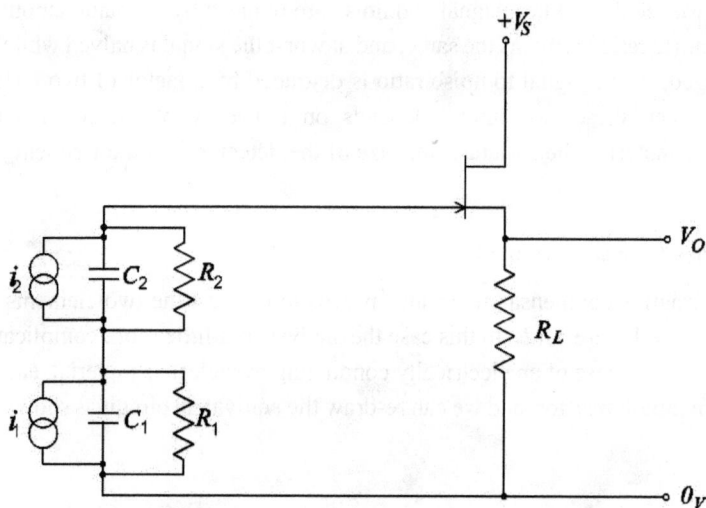

Figure 5.13 Series compensated detector with individual bias resistors.

The observed signal at the input of the amplifier is the sum of the voltages generated across the two elements, so we have:

$$v = i_1 Z_1 - i_2 Z_2 \tag{5.17}$$

$$\text{where } Z_1 = \frac{R_1}{1+j\omega R_1 C_1} \text{ and } Z_2 = \frac{R_2}{1+j\omega R_2 C_2}$$

For v to be zero we must therefore have:

$$\frac{i_1 R_1}{(1+j\omega R_1 C_1)} = \frac{i_2 R_2}{(1+j\omega R_2 C_2)} \tag{5.18}$$

Substituting from equation 5.2, this becomes:

$$\frac{pA_1 R_1}{(1+j\omega R_1 C_1)} \frac{dT}{dt} = \frac{pA_2 R_2}{(1+j\omega R_2 C_2)} \frac{dT}{dt} \tag{5.19}$$

Now, for the electrically conducting pyroelectric elements we have:

$$C_1 = \frac{\varepsilon_0 \varepsilon_r A_1}{d_1}, \qquad C_2 = \frac{\varepsilon_0 \varepsilon_r A_2}{d_2} \tag{5.20}$$

and:

$$R_1 = \frac{\rho d_1}{A_1}, \qquad R_2 = \frac{\rho d_2}{A_2} \tag{5.21}$$

from which we can see that:

$$R_1 C_1 = R_2 C_2 = \varepsilon_0 \varepsilon_r \rho \tag{5.22}$$

Combining equations 5.19, 5.21, and 5.22 we arrive at:

$$d_1 = d_2 \tag{5.23}$$

This means that, for effective compensation in the series arrangement, the only requirement is that the thicknesses of the two elements are the same. This is generally easily arranged because the two elements are usually formed on the same pyroelectric chip. The areas of the two elements do not need to be equal because both the capacitance and pyroelectric charge generated scale with the area of the element and balance out.

5.3.2.1. Responsivity and Noise for a Series Compensated Detector

In the case of a series compensated detector, we replace the equivalent circuit shown in Figure 5.11 with that shown in Figure 5.14.

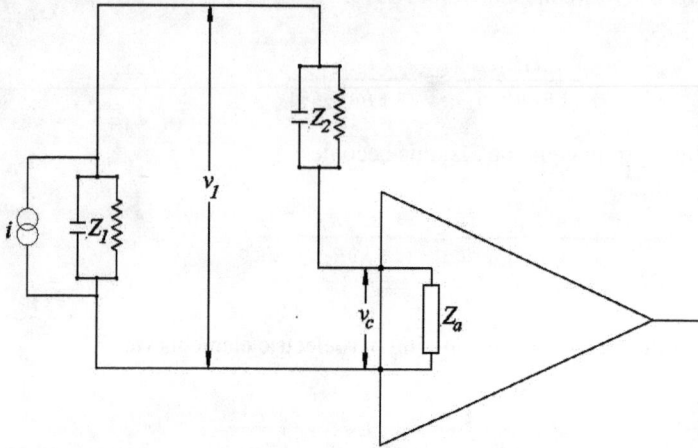

Figure 5.14 Equivalent circuit for series compensated detector.

For the uncompensated detector the voltage seen by the amplifier is, as before, given by equation 5.7. For the series compensated detector, however, the combined impedance, Z_c, is given by:

$$\frac{1}{Z_c} = \frac{1}{Z_1} + \frac{1}{Z_2+Z_a} \tag{5.24}$$

so

$$Z_c = \frac{Z_1(Z_2+Z_a)}{Z_1+Z_2+Z_a} \tag{5.25}$$

The current i flows through this combined impedance and generates the voltage

$$v_1 = \frac{iZ_1(Z_2+Z_a)}{Z_1+Z_2+Z_a} \qquad (5.26)$$

However, the amplifier sees the voltage, v_c, given by:

$$v_c = \frac{v_1 Z_a}{Z_2+Z_a} \qquad (5.27)$$

so we have:

$$v_c = \frac{iZ_1(Z_2+Z_a)Z_a}{(Z_1+Z_2+Z_a)(Z_2+Z_a)} \qquad (5.28)$$

i.e.

$$v_c = \frac{iZ_1 Z_a}{(Z_1+Z_2+Z_a)} \qquad (5.29)$$

From equations 5.7 and 5.29 we can again calculate the ratio of compensated to uncompensated voltages seen by the amplifier:

$$\frac{v_c}{v_u} = \frac{Z_1+Z_a}{Z_1+Z_2+Z_a} \qquad (5.30)$$

If the amplifier has a very high input impedance, so that Z_a is much larger than both Z_1 and Z_2, this ratio approaches 1 and the responsivity of the series compensated detector is equal to that of the uncompensated detector. This will generally be true for large area detectors.

For small detector elements, however, the value of Z_a is likely to be similar to Z_1. If the active and compensating elements are of the same size, so that the three impedances are approximately equal, then $v_c/v_u \approx 2/3$. However, if we make the compensating element larger than the active element, so that its capacitance increases and its resistance decreases, Z_2 will become smaller than Z_1 and Z_a, and the ratio will tend back to 1.

So, we see that, in the case of series compensation, we can achieve a similar responsivity to the equivalent uncompensated detector by making the compensating element large compared with the active element, even if the amplifier input impedance is not significantly larger than the active element impedance.

The total noise in a series compensated detector is simply the noise of the compensating element added, in series, to the noise of the active element. If the two elements are of the same size, the total RMS noise is therefore increased by $\sqrt{2}$ (unless amplifier voltage noise is dominant, in which case it is unchanged). If, however, the compensating element is significantly larger than the active element, R_2 is much smaller than R_1 and C_2 is much larger than C_1, so the second third and fourth terms of equation 3.50 become significantly smaller, and the compensating element adds a negligible amount to the RMS noise, unless thermal noise or amplifier voltage noise is dominant.

So, for a series compensated detector, the best case is that the responsivity is unchanged compared with the uncompensated equivalent, and the noise is increased by a small or insignificant amount, so the signal to noise ratio not degraded. The worst case is that the responsivity is reduced by 2/3 and noise is increased by $\sqrt{2}$, so that the signal to noise ratio is degraded by a factor $\sqrt{2}/3$. As in the case of a parallel compensated detector, the performance of a specific detector will be somewhere between these two limits, depending on the properties of the detector and the operating frequency. A series compensated detector with a large compensating element will generally have a performance very close to the equivalent uncompensated detector, with the added advantage of relative immunity to ambient temperature variations.

5.4. Dual and Quad Element Detectors

A single element pyroelectric detector can be used, with suitable optics, to detect a moving target that has a different temperature to the background. It will not, however, give any information about the size of the target or its direction and speed of movement.

Much more information can be derived from a dual element detector, which has two active elements in the same package, viewing the target through the same optics. Figure 5.15 shows an example of a dual element detector, each active element (the black rectangles) having its own compensating element (the gold rectangles). The active elements have a black coating so that they absorb the incident radiation whereas the compensating elements have a gold reflecting surface so that they do not absorb radiation. Each active-compensating element pair has its own FET, bias resistor, and load resistor, so this detector has two independent outputs.

Figure 5.15 A dual element compensated detector.

If the image of a moving target, such as a man, is of similar width to that of the active elements, and moves across this detector from left to right, we see two independent outputs of the form illustrated in Figure 5.16. The output signals are here plotted as a function of position of the image, along the horizontal axis. The form is the same when plotted against time for the moving image.

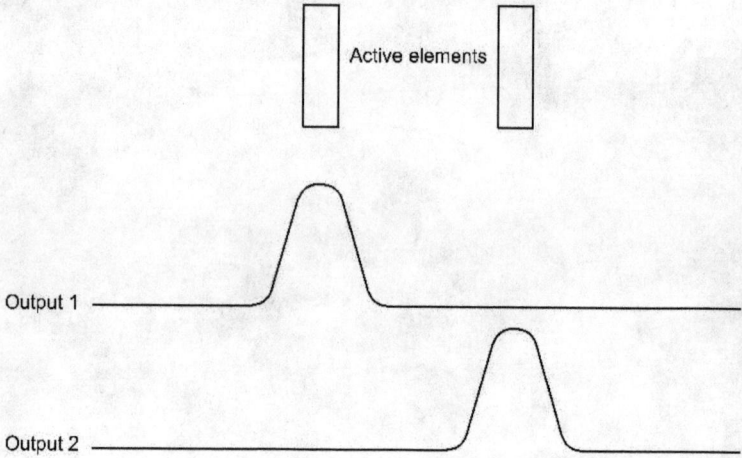

Figure 5.16 Output signals from dual element detector as image moves across.

Analysis of these two output signals as a function of time enables the determination of the width of the image and its velocity (speed and direction of movement). With knowledge of the focal length of the lens or other optics producing the image, this information can then be related to the angular size and velocity of the moving object. Determination of the linear size and velocity of the object requires knowledge of its range (distance from the lens to the object), but this can probably be deduced from the layout of the scene being viewed.

Another common arrangement in intruder alarms is a single element compensated detector with both elements exposed and absorbing radiation. In this case there is a single output of the form shown in Figure 5.17. For a moving source that is warmer than the background, the output signal is positive when the image crosses one element and negative when it crosses the other element. It is therefore possible with this arrangement to extract almost as much information about the source as with the dual compensated arrangement of Figure 5.15. It is not possible, however, to distinguish between a warm object moving from left to right and a cold object moving right to left.

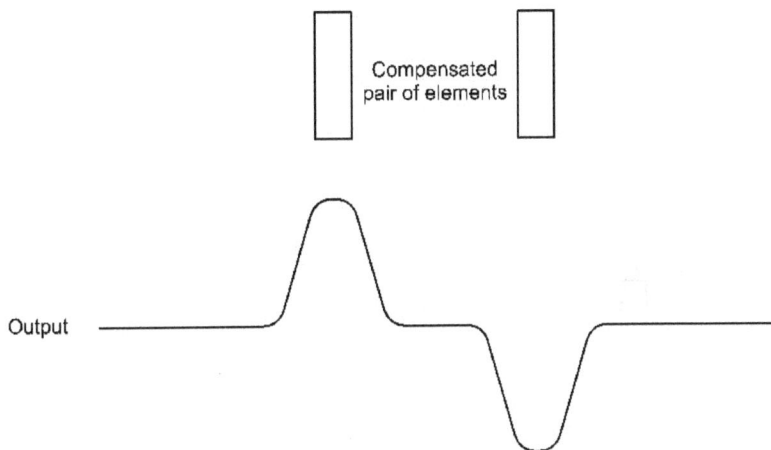

Figure 5.17 Output signal from a compensated pair of elements.

A further refinement is to have two compensated pairs, usually with square elements, arranged in a square format, as illustrated in Figure 5.18. Here the + and – signs indicate the polarization direction of each element. An arrangement like this is usually termed a quad element detector.

Movement of an image from left to right or from top to bottom of this detector will produce similar signals, but by the judicious arrangement of the four elements in two compensated pairs it is possible to gain information about the movement of an object within a two-dimensional plane.

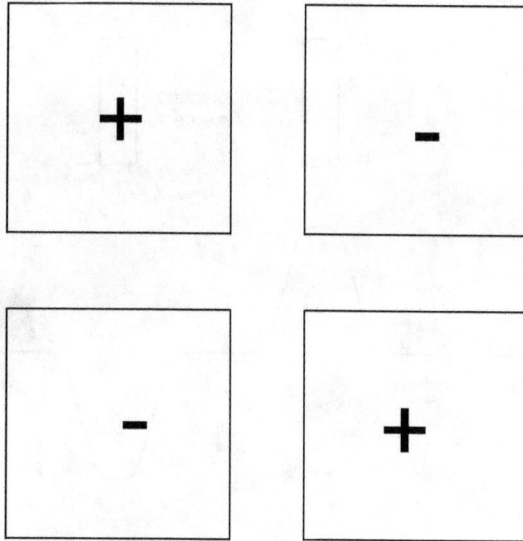

Figure 5.18 Quad element detector arrangement.

5.5. Multi-element Arrays

Single, dual, and quad element detectors are very useful for detecting the presence of moving objects, and for other sensing applications. However, there are many applications for which more information on the spatial characteristics of a scene is required. This may simply be to obtain unambiguous information about the size, shape, and direction of movement of an object being detected, or it may be the requirement for a full thermal image of the scene. Over the last 30 or 40 years a range of one- and two-dimensional pyroelectric detector arrays have been developed with these requirements in mind.

5.5.1. Linear arrays

One example of a linear array is shown in Figure 5.19. This is a 64-element linear array developed during the 1980s by The Plessey Company. It has 64 discrete elements defined on a single chip of pyroelectric ceramic by photolithography; after which the elements are thermally separated by saw cuts between them. The package includes 64 JFET chips buffering the 64 separate outputs.

Figure 5.19 64 element linear array.

This array was developed for "push-broom line scanning" in which the array looks vertically downwards, through a suitable lens, from an aircraft as it passes over an area of which a thermal image is required. The aircraft moves in a direction perpendicular to the long axis of the array, so that the array sweeps across the image of the ground and produces an image 64 pixels wide. The aircraft flies back and forth across the scene building up an image with a succession of contiguous strips.

5.5.2. Two dimensional arrays

For real time thermal imaging, however, a two-dimensional array is needed. An early example of this is shown in Figure 5.20. This is a 6 by 6 element two-dimensional array, with 36 separate outputs. It was developed, again by the Plessey Company, for prototype Infrared Terminally Guided Submunitions (TGSM), but was never put into production.

Figure 5.20 6x6 element two-dimensional array.

The use of a separate buffer amplifier and output lead for each element obviously limits the possible number of elements in a two-dimensional array. For this reason, significant effort was applied to the development of technologies enabling the multiplexing of the outputs from arrays within the package. One of the first examples of such a hybrid array is shown in Figure 5.21. This consists of a square chip of pyroelectric ceramic material on which 256 elements are defined by photolithography; this chip is then bonded to a CCD readout chip that has 256 inputs multiplexed into a single output. An example of the quality of thermal image that can be obtained using interpolation between the elements of such an array is shown in Figure 5.22. This is a thermal image of the author.

Figure 5.21 Hybrid 16x16 array.

Figure 5.22 Thermal image obtained using a 16x16 element array with interpolation.

5.5.3. Read-Out Integrated Circuits

In order to make a two-dimensional hybrid pyroelectric array, we need an Application Specific Integrated Circuit, or ASIC, that is matched to the proposed detector array. For example, the 16x16 element hybrid array shown above has 256 elements arranged in a square matrix on a 100µm pitch. This requires a matching ASIC that has 256 input pads in the same square format on the same pitch. An example of such a device is shown in Figure 5.23. This example was developed in the 1980s and is a Charge Coupled Device (CCD) multiplexer; later hybrids used CMOS multiplexers because these generally give a better signal to noise performance.

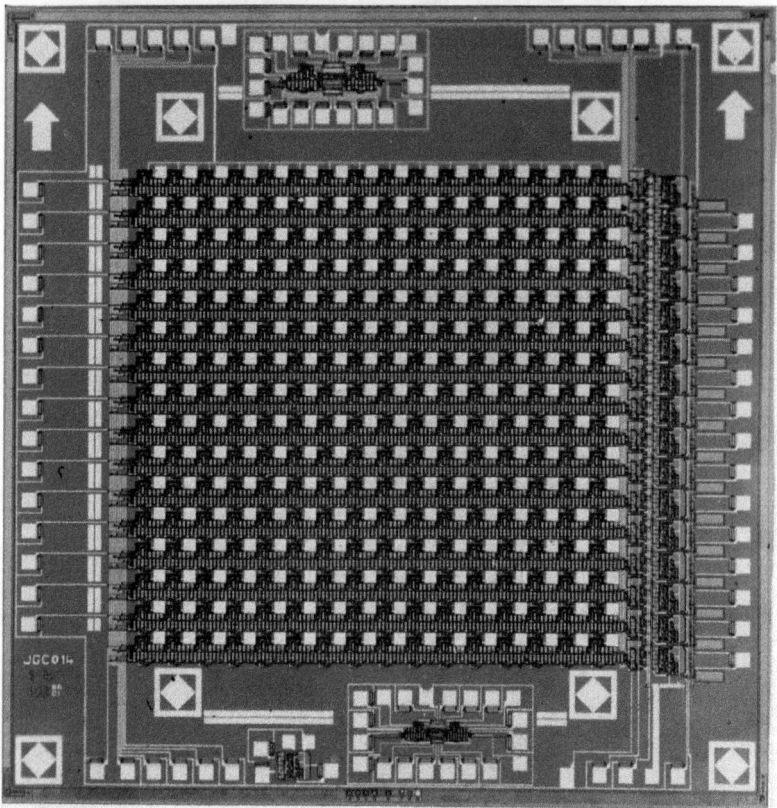

Figure 5.23   256 input CCD multiplexer

In order to connect the pyroelectric array to the readout ASIC, a specialised solder bonding technique was developed, by means of which separate solder bumps were formed on each of the input pads of the ASIC. The pyroelectric array, with suitably processed matching electrodes on its under-surface, was solder bonded to the ASIC, forming 256 mechanical and electrical joints.

A lower cost technology has, more recently, been developed, in which the 256 contacts are made using a conducting epoxy instead of solder. This technology has been used to manufacture the arrays used in the people counters and low cost thermal imagers described in Chapter 6.

5.6.   The Pyroelectric Vidicon

The discussion of pyroelectric devices would not be complete without mentioning the pyroelectric vidicon. This was developed by English Electric Valve Co. Ltd. (now E2V) and the Royal Signals and Radar Establishment (RSRE), Malvern (now Qinetiq) in the 1970s [38] This used a standard vidicon camera tube but replaced the normal photoconductive target with a wafer of pyroelectric material. Quite good thermal images were obtained, and it was the best available un-cooled thermal imager for several years.

The vidicon was a vacuum tube imaging device that was developed in the 1950s. It was widely used in television cameras until the 1980s, when they were superseded by solid state CCD and CMOS sensors. Figure 5.24 is a schematic diagram of a vidicon tube, indicating the mode of operation. An electron beam is emitted from a cathode and is accelerated by a series of grids towards a target held at a positive potential with respect to the cathode. The electron beam is aligned, focussed, and scanned across the target in raster fashion by means of a set of coils around the outside of the vidicon tube.

The target comprises a glass plate, sealed onto the end of the vacuum tube, which is coated on the inside, first with a transparent conductive layer, and then with a photoconductive layer. A video signal current is read out of a connection to the conductive film. An image of a scene is projected through the glass faceplate and conductive film onto the photoconductive layer. The resistivity of the photoconductor varies according to the brightness of the image formed upon it, and this modulates the readout current as the electron beam is scanned across the target.

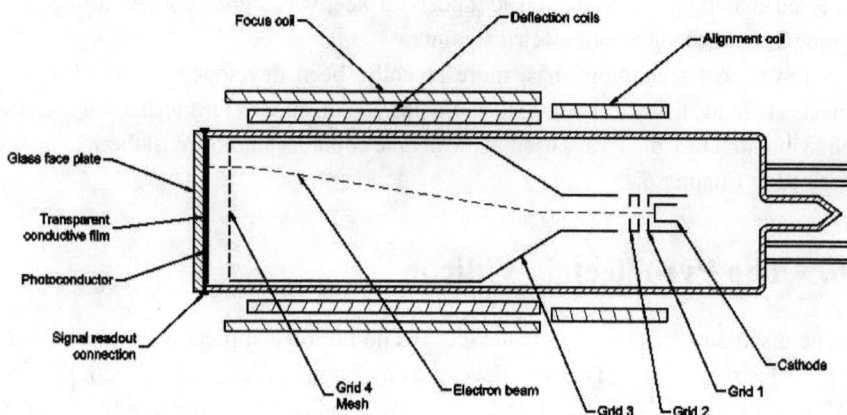

Figure 5.24 Standard vidicon camera tube.

The pyroelectric vidicon uses the standard vidicon tube with the glass faceplate and photoconductor replaced by a germanium faceplate and a wafer of pyroelectric material, such as TGS, as shown in Figure 5.25. A thermal image is projected through the germanium faceplate onto the pyroelectric by means of a germanium lens.

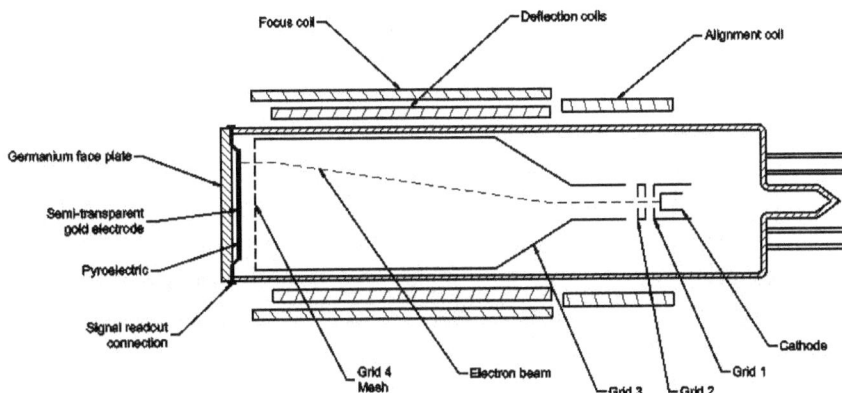

Figure 5.25 Pyroelectric vidicon.

Being a thermal detector, the pyroelectric wafer cannot be mounted directly on the germanium faceplate, as this would form a good heat sink and severely compromise the pyroelectric signal. A special mount is therefore used, which suspends the pyroelectric wafer by its edges with a small gap between the pyroelectric and the faceplate, thereby providing thermal isolation.

A further problem with the pyroelectric vidicon is that, as the thermal scene is modulated, either by chopping or panning the image, alternate positive and negative charges are generated on the back surface of the pyroelectric. The electron beam, however, can only neutralize positive charges, so some means of neutralizing the negative charges is necessary in order to obtain a continuous video signal. The normal method of achieving this, when using a TGS target, was to use a soft vacuum (i.e. leaving some residual gas) so that a positive ion beam is formed to balance the negative electron beam. Some development was done on the alternative approach of using a conductive ceramic pyroelectric wafer in a hard vacuum, but the requirement for this was superseded by the development of solid-state un-cooled thermal imagers.

6. Applications of Infrared Detectors

6.1. Introduction

In this chapter we present a picture of the range of applications for pyroelectric infrared detectors, ranging from simple single or two element devices as used in intruder alarms to multi-element pyroelectric arrays for thermal imaging and radiometry.

6.2. Sensors

Because pyroelectric detectors give an output signal only when there is a change in incident radiation, they are particularly suited to the detection of moving warm objects, or other phenomena that produce a thermal change in a scene. The most common example of this is the detection or monitoring of moving people, and the largest markets for pyroelectric detectors at the present time are intruder alarms and people counters.

6.2.1. Intruder Alarms

The first commercial application of pyroelectric detectors was intruder alarms. A human being is usually at a higher temperature than the background, and as he or she moves around the thermal scene will change. For a pyroelectric detector to see this change, an optical system must be engineered such that the thermal image of the person moves on and off the detector as the person moves. This is commonly achieved by means of an array of lenses or mirrors, as illustrated in Figure 6.1. In this example an array of seven infrared lenses is used to focus parts of a scene onto a single element pyroelectric detector.

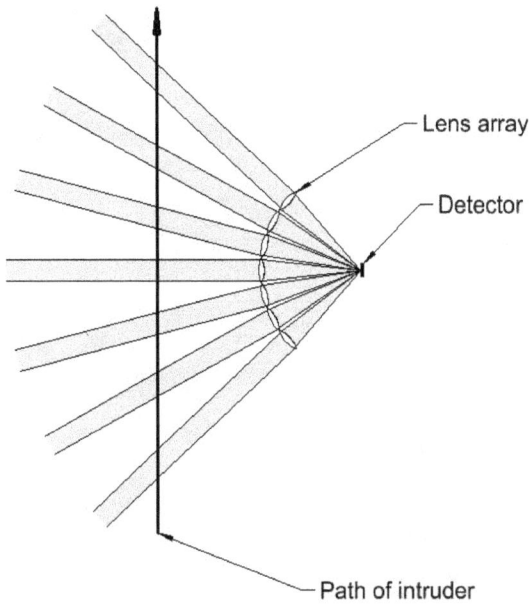

Figure 6.1 Principle of operation of pyroelectric intruder alarm.

The shaded areas in the diagram indicate the parts of the scene focused onto the centre of the detector. As an intruder walks across the scene, as indicated by the long arrow, he moves repeatedly into and out of the field of view of the detector, and so produces an alternating signal.

There are now many manufacturers of so-called passive infrared (PIR) intruder alarms using pyroelectric detectors, the majority of which use a plastic Fresnel lens array with a dozen or more facets. A typical example of an intruder alarm sensor is shown in Figure 6.2, and the lens array from this is shown in Figure 6.3. This example has 11 long rectangular facets in the upper two thirds of the lens array; these provide the major segmented field of view, looking out from the detector a few degrees below the horizontal. Below these are another 8 facets, covering the part of the field of view at approximately 20° below the horizontal, as shown schematically in Figure 6.4. Below these again are a further 5 facets, covering the space nearer to the detector.

Figure 6.2 A typical PIR intruder alarm sensor.

Figure 6.3 The Fresnel lens array used in the PIR sensor above.

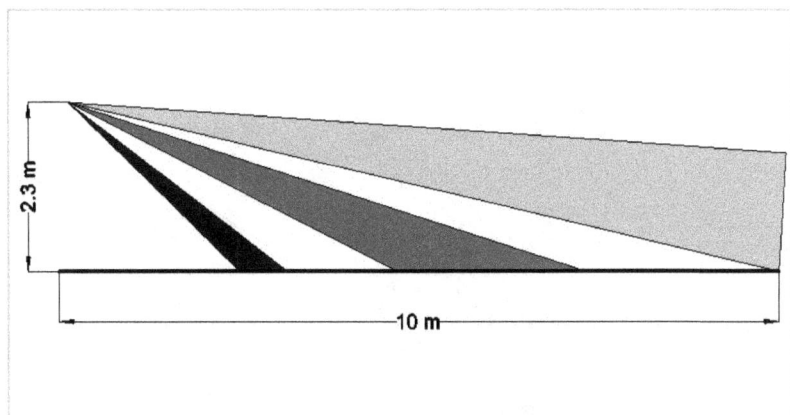

Figure 6.4 Side elevation of PIR fields of view.

These lens arrays are usually used with a compensated pair of detector elements so that, as a person walks through each zone, a positive signal followed by a negative signal (or vice-versa) is generated. Thus, in plan, the active fields of view comprise 24 pairs of zones, as illustrated in Figure 6.5, each pair producing a positive and negative signal as a person passes through.

A more sophisticated approach is to use a quad element detector, as described in section 5.4, to give improved resolution in the vertical plane as well as the horizontal plane. This improves the sensitivity of the alarm when the intruder is moving towards or away from the detector.

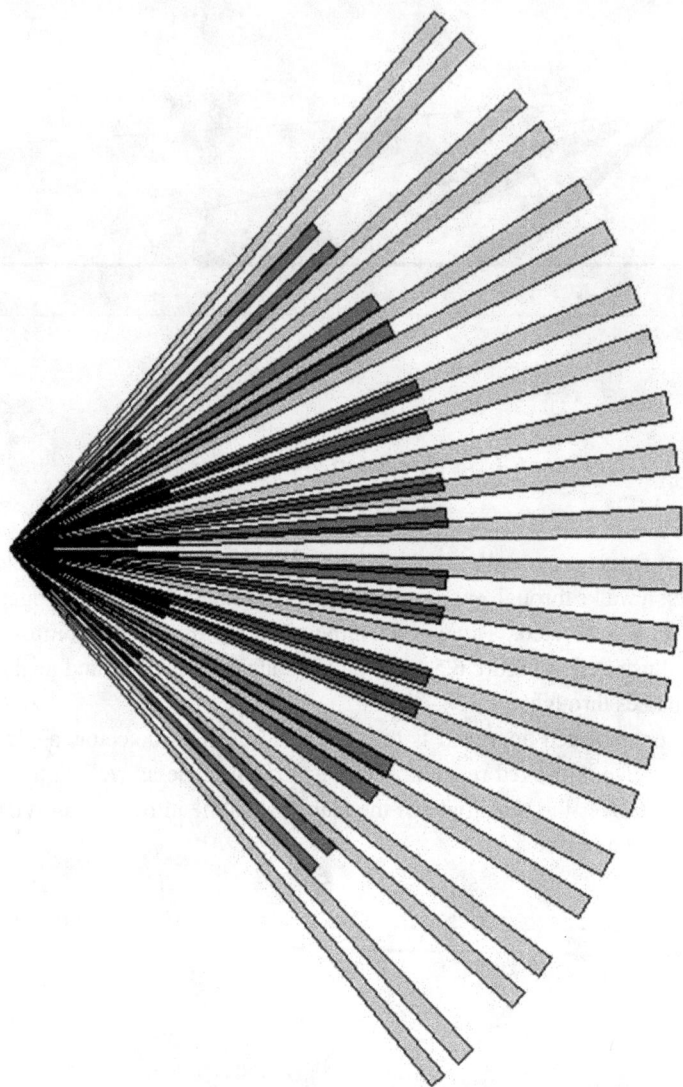

Figure 6.5 Plan of PIR fields of view.

6.2.2. Light Switches and Occupancy Sensors

Units very similar to intruder alarms can be used to control so-called security lights. These are mounted on the outside of houses and other buildings in such a position as to illuminate the footpath or driveway when a person or vehicle approaches the house.

A similar application is occupancy sensors. In this case, the requirement is to determine whether or not a room is occupied so that lights, extraction fans, or other services can be switched on when people enter the room and switched off again a specified time after they have left. These units are typically mounted on the ceiling at or near the centre of the space to be monitored, looking vertically downwards. The field of view is divided into zones by a plastic lens array, in similar manner to an intruder alarm, but in this case the zones are usually arranged in a circular pattern centred on a spot vertically below the sensor. A typical unit is illustrated in Figure 6.6, and the inside of a typical lens array is shown in Figure 6.7.

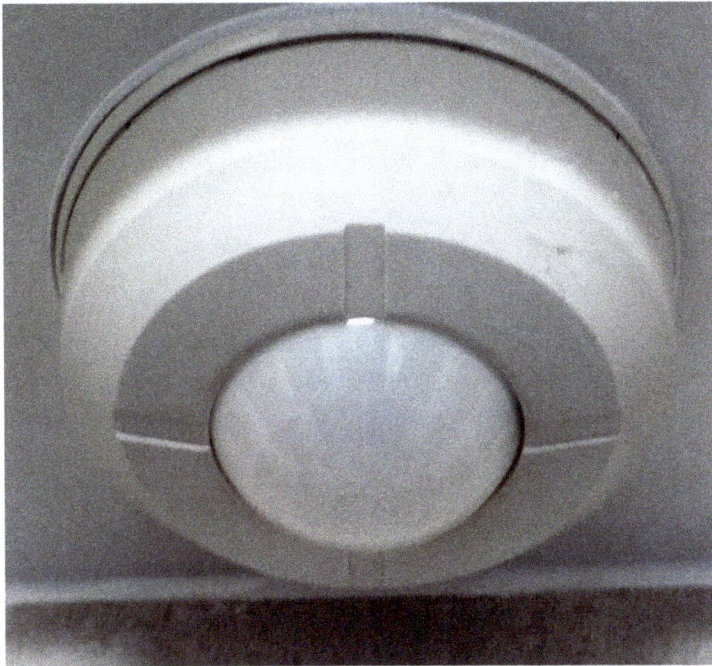

Figure 6.6 A typical occupancy sensor.

Figure 6.7 A typical occupancy sensor lens array.

Light switches and occupancy sensors usually include a light level adjustment, so that a light can be switched on only at night, and a hold time adjustment, so that a light or fan can be kept on for a specified time after the area or room has been vacated.

6.2.3. People Counters

Intruder alarms and occupancy sensors work well at detecting the presence of a moving object within their field of view, but they are not able to tell how many objects there are, what sizes they are, where they are within the field of view, or in what direction they are moving. Most, if not all, of this information is required in order to count people within a given space or moving past a certain point.

What is needed to count people (or other objects), is not a single or dual detector and a lens array, but an array of detectors and a single lens. By imaging a scene with a single infrared lens onto an array of detectors and monitoring the signals from each of the detector elements within the array, the position and direction of movement of objects within the scene can be determined.

People counters are now available using low-cost pyroelectric detector arrays, typically of 256 elements arranged in a 16x16 matrix, and low cost, single element, germanium lenses. Sophisticated software algorithms have been developed to process the 256 multiplexed signals from the detector elements so that people can be detected and tracked within the overall field of view. This allows, for example, the simultaneous counting of people going in and out through a doorway.

One major application for this technology is in supermarkets, where pyroelectric counters are used to count people waiting in the check-out queues, as well as those entering and leaving the store. Integrating this data with the information from the check-out tills enables the store manager to deploy his staff in the optimum manner to provide the best customer service.

6.2.4. Flame Detectors

The majority of domestic, and many industrial, fire alarm systems use smoke detectors as their primary sensors. In specialist applications, such as fuel storage depots and oil rigs, where a hydrocarbon fire is likely to produce flame immediately, it is better to use flame detectors rather than smoke detectors.

One characteristic of a hydrocarbon flame is a strong emission of infrared radiation from hot carbon dioxide at wavelengths of 2.7 μm, 4.3 μm, and 15 μm. This radiation is modulated at the characteristic flicker frequency of the flame, which is normally within the range 1 Hz to 20 Hz, which is ideally matched to the maximum sensitivity of a pyroelectric detector. Atmospheric CO_2 will absorb radiation of the same wavelengths, but the emission bands of the hot CO_2 in the flame will be broader than the absorption bands of the cold atmospheric CO_2. A typical flame detector, therefore, uses a pyroelectric detector fitted with a narrow band filter centred at 4.3 μm, with the bandwidth of the filter chosen so that the detector will see the 'shoulders' of the emission band on either side of the narrower absorption band. To minimise false alarms caused by moving or modulated hot objects in the scene, a second "guard channel" detector is included. The guard detector is fitted with a filter at a slightly different wavelength, so that it will see moving hot objects, but it will not see the hot CO_2. The difference between the two signals is used to trigger an alarm.

Some high-performance flame detectors use a small array, similar to that used in people counters, and a sapphire lens, so that the position and turbulence characteristics of the flame can be identified.

6.3. Monitors and Radiometers

The sensing applications described above use the inherent property of a pyroelectric detector to respond only to changes in incident radiation. There are, however, various monitoring and measuring applications in which a constant, stationary source is deliberately modulated at a specific frequency. A selection of monitoring and measuring applications is described below.

6.3.1. Infrared Spectral Analysis

6.3.1.1. Non-dispersive Gas Analysis

Carbon dioxide absorbs radiation in a narrow band centred on 4.3 μm. Other gasses absorb at their own characteristic wavelengths, as listed for some important gases in Table 6.1. Pyroelectric detectors can therefore be used for gas analysis by introducing the gas between a modulated broad band infrared source and a pyroelectric detector fitted with an appropriate narrow band filter. This technique is known as non-dispersive infrared analysis.

Gas	Characteristic absorption wavelength (μm)
CO_2	4.26
CO	4.69-4.79
SO_2	7.35
CH_4	3.31
C_6H_{14}	3.39
NO	5.26
NO_2	6.17
N_2O	4.50

Table 6.1 Characteristic absorption wavelengths for a selection of gases.

To improve the sensitivity and accuracy of this type of analysis, the infrared light is split, and passes along two similar paths, one containing the gas to be analysed, the other being a reference path without the gas. Comparison is then made by switching the detector alternately between the two channels.

The greatest commercial success in this area involves the measurement of the concentration of carbon dioxide in the air. The sensitivity vs. cost tradeoff in this application is optimum for pyroelectric devices. Carbon dioxide is an unseen poison that can make beer-cellars, abattoirs, and breweries dangerous places. In medical care "capnography" monitors the gas concentration over the course of a patient's breathing cycle, and pyroelectric devices offer the tantalizing possibility of doing this remotely.

There are several unexploited possibilities for using pyroelectric arrays: not only for different gases but also to image carbon dioxide in the atmosphere at different wavelengths in order to gauge temperature and turbulence at different distances. So-called CAT (Clear Air Turbulence) is a serious aeronautical problem.

6.3.1.2. Dispersive Spectral Analysis

Dispersive analysis also uses a broad band infrared source, but in this case the radiation is dispersed into a continuous spectrum by means of a prism or diffraction grating. A detector then views this spectrum through the sample to be analysed and is scanned slowly through the spectrum. Thus, a complete absorption spectrum for the sample is produced. With this technique, the intensity of radiation at each wavelength is low, and decreases as the wavelength resolution increases. Very sensitive detectors are therefore necessary, and most manufacturers use TGS detectors, despite their disadvantages.

6.3.2. Laser Monitors

Carbon dioxide lasers produce powerful beams of infrared radiation at a wavelength of 10.6μm. Pyroelectric ceramic detectors are ideal for monitoring these laser beams because they are robust, both mechanically and thermally, and the ceramic material absorbs reasonably well at 10.6μm. There are essentially three types of measurement that a laser operator may wish to make:

6.3.2.1. Laser Power Monitoring

A ceramic detector with face electrodes and typical amplifier for low frequencies, as shown in Figures 5.7 and 5.1, may be used for monitoring the average power in a laser beam. However, if absorption takes place in the front electrode or an absorbing coating, it is likely that this will get too hot and be damaged or evaporated. For this reason, it is normal to use a semi-transparent front electrode, such as a thin layer of nichrome (an alloy of nickel and chromium), so that most of the incident radiation passes through the electrode and is absorbed in the ceramic itself.

Using this arrangement with a 0.5 mm thick ceramic element will typically absorb about 75 % of the incident energy, and if the detector is 5 mm square and the ceramic has a relative permittivity of about 300, it will have a capacitance of about 140 pF. Using a bias resistor of 20 GΩ will give an electrical time constant of about 3 s, so that modulation of the laser beam at 10 Hz or more means that the detector operates on the 1/f part of its responsivity spectrum. The detector will therefore need to be calibrated at a specific modulation frequency before use.

Although a pulsed laser provides its own modulation so that a pyroelectric detector will respond, to measure the power in a CW laser it must be modulated at a suitable frequency by, for example, a mechanical chopper. For very high-power lasers, it will probably be necessary to use a beam splitter to extract a known fraction of the total energy for measurement.

6.3.2.2. Pulse Energy Measurement

In the case of a pulsed laser, the operator may wish to measure the total energy within each pulse. To achieve this, a ceramic detector with a semi-transparent front electrode is again used, but the value of the bias resistor is reduced so that the electrical time constant of the detector is lower than the time between pulses, but long compared with the pulse length. The detector will then integrate the incident power over each pulse to give a measure of the total energy in the pulse.

For example, if the 5 mm square 0.5 mm thick detector described above is used with a 180 Ω bias resistor, the detector time constant will be about 25 ns and the response of the detector to a series of 20 ns laser pulses at a repetition rate of 5 MHz will be as shown in Figure 6.8. The detector effectively integrates the laser pulse power, and the amplitude of the output is proportional to the energy (power times time) of the pulse.

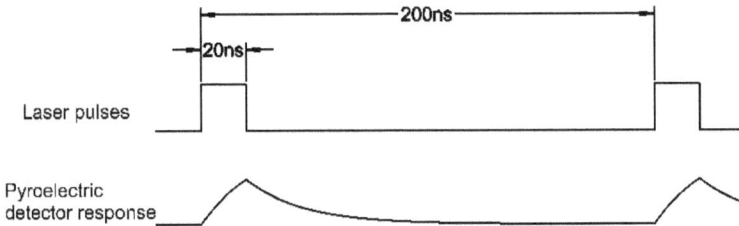

Figure 6.8 Response of detector with 25ns electrical time constant.

6.3.2.3. Pulse Shape Monitoring

In order to follow the shape of a laser pulse, a detector must have a flat frequency response up to the highest frequency of interest. This means that the electrical time constant must be short compared with the width of the pulse to be monitored.

To achieve this, a 5 mm square detector 0.5 mm thick can again be used, but in this case, as described in section 5.2.2, edge electrodes would be used (on two opposing 5 mm x 0.5 mm faces), but the laser beam would still be incident on the 5 mm square face. The capacitance of this arrangement would be approximately 1.4 pF. A bias resistor of 50 Ω (chosen to match the impedance of a standard transmission cable) then gives an electrical time constant of about 70 ps. This will follow a 2 ns laser pulse reasonably well, as shown in Figure 6.9. Obviously, the responsivity will be very low with such a low value bias resistor but is adequate for monitoring high powered laser pulses.

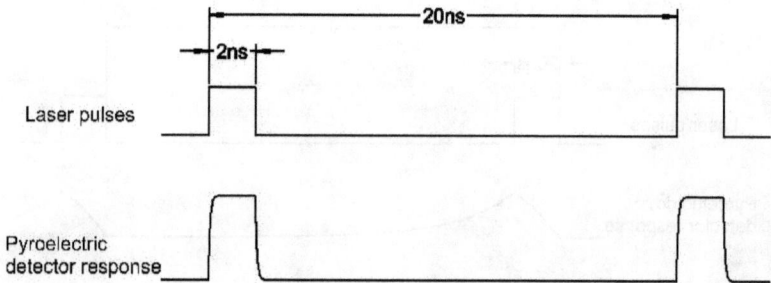

Figure 6.9 Response of detector with 7 ps electrical time constant.

6.3.3. Radiometers

If it is necessary to know the temperature of an object, this can often be achieved with the aid of a thermometer, thermocouple, thermistor or other temperature sensor in contact with the object.   However, there are situations in which a non-contact temperature measurement is required.   A pyroelectric radiometer may be used for this purpose.

A radiometer can be made by arranging a pyroelectric detector with a suitable chopper and optics (either a lens or a mirror) so that the image of the object to be measured fills the field of view of the detector, as shown in Figure 6.10.

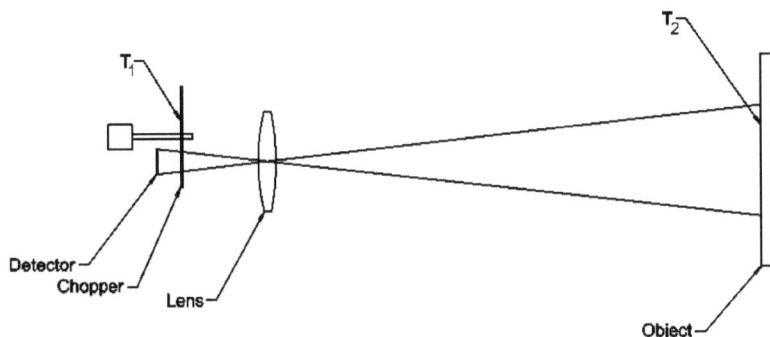

Figure 6.10 Schematic arrangement of a radiometer

As the chopper rotates the detector will alternately see the object to be measured, at temperature T_2, and the chopper, at temperature T_1. The amplitude of the signal generated by the detector will be proportional to the difference in radiation received from the object and the chopper.

If both the object and the chopper are black body radiators, then the radiation from each is of the form shown in Figure 3.9 for the appropriate temperature. The total radiation emitted from each body is the area under this spectral radiance curve, and performing the integration yields the Stefan-Boltzmann law:

$$M = \left(\frac{2\pi^5 k^4}{15 h^3 c^2}\right) T^4 \qquad \qquad 6.1$$

in units of Wm^{-2}, where M is the total radiated power per unit area, or radiant emittance. Thus, we see that the radiant emittance is proportional to T^4 and, if all the radiation from the two sources fell on the detector, the signal amplitude would be proportional to $T_2^4 - T_1^4$.

However, the radiation is restricted by the pass band of the infrared windows and lenses between the sources and the detector, so this proportionality is only approximate. To obtain an accurate measure of temperature, therefore, a detailed calibration matrix must be generated covering the full range of T_1 and T_2 over which the radiometer is required to operate.

One interesting application of pyroelectric radiometers is in the Stratospheric Sounding Unit (SSU) used on the NOAA polar orbiting satellites to measure stratospheric temperatures between 1978 and 2006 [39]. In this instrument the rotating chopper was replaced by a pressure modulated cell of CO_2 gas. By selecting a particular mean pressure in the CO_2 cell, the temperature of atmospheric CO_2 at an altitude corresponding to that pressure could be measured. Each instrument contained three radiometers to measure the temperature of CO_2 simultaneously at three different altitudes. The data gathered can potentially provide valuable information on global temperature trends.

6.4. Thermal Imaging

If we take the arrangement shown in Figure 6.10, but substitute a detector array for the single element detector shown in the figure, we have the basis of a thermal imager. To produce the thermal image, the output signals from the elements of the detector array must be fed out of the imager as a serially multiplexed signal. This signal is then used to generate an image on a suitable display.

As mentioned in section 5.5, two-dimensional pyroelectric hybrid arrays of 16x16 elements were developed in the 1980s using a CCD readout multiplexer. Since then, various pyroelectric arrays, suitable for thermal imagers, have been produced using CMOS readout chips with element counts up to 4880. An example of a low-cost thermal imager, using a 16x16 element pyroelectric array, is shown in Figure 6.11 viewing a hot radiator below a cold window.

Figure 6.11   Example of low-cost thermal imager.

6.4.1. Modulation

The elements of a two-dimensional array are read out sequentially, line by line from the top to the bottom of the image plane. For the optimum performance, we therefore need to modulate the incident radiation by means, for example, of a mechanical chopper moving from top to bottom of the array in synchronism with the electrical readout.

The simplest form of chopper is a sectored disc, as illustrated in Figure 6.12. With this arrangement, however, there is a phase difference in the modulation of the elements at the right-hand side of the array compared with the elements at the left-hand side. This will result in a non-uniform signal from the array when viewing a uniform temperature.

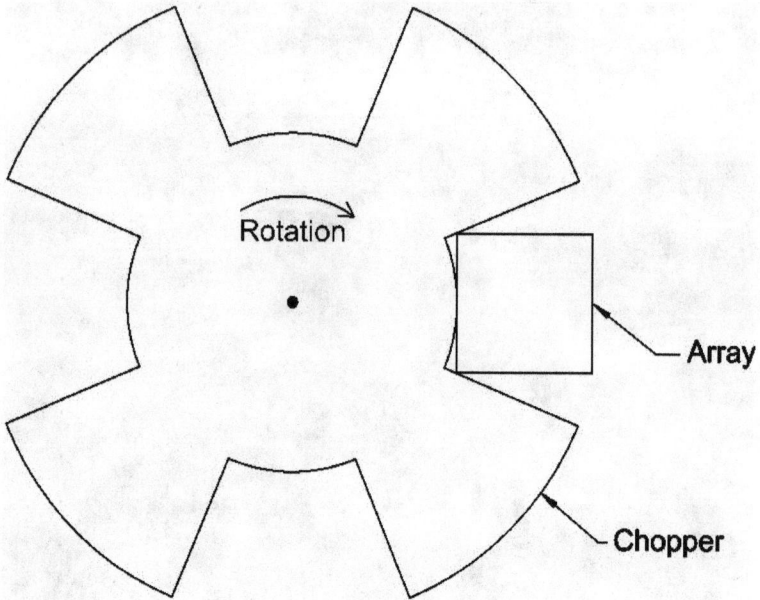

Figure 6.12 Sectored chopper.

The performance can be improved by using chopper apertures with spiral edges, as illustrated in Figure 6.13. Here the edge of the chopper aperture moves down the array in linear fashion, but there is a residual phase difference along a horizontal row of elements due to the curvature of the chopper edge.

A further improvement can be achieved by reducing the number of apertures in the chopper blade. Figure 6.14 illustrates a spiral chopper with only one aperture. In this case, the edge of the chopper aperture has a lower curvature so the phase variation along a row of elements is lower, and the resulting signal is more uniform. The single bladed chopper has an additional advantage in that the motor operating the chopper can go faster for a given chopping frequency, which will generally give more stable operation. It does, however, have the disadvantage that the chopper blade is no longer symmetrical, and will therefore need to be balanced by adding weights in appropriate places.

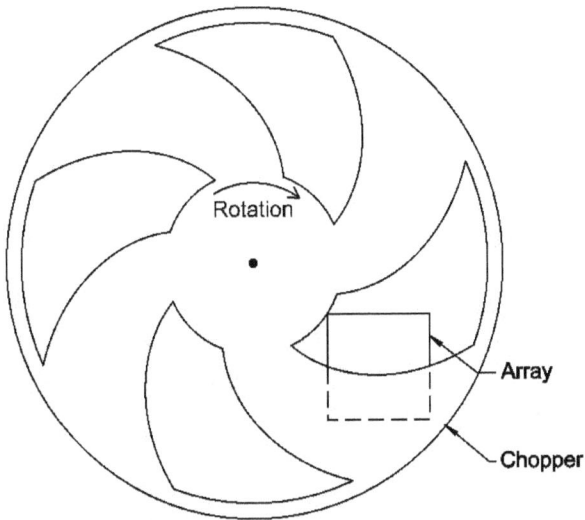

Figure 6.13 Spiral chopper with four apertures.

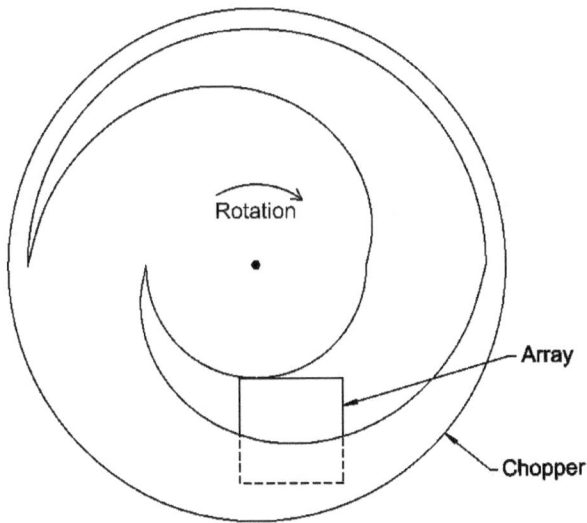

Figure 6.14 Single aperture spiral chopper.

6.4.2. Image processing

The multiplexed signal from the modulated array is in the form of sampled voltages, one for each element of the array, successive fields of these sample voltages being for the chopper open and the copper closed states. In order to generate the appropriate signal to apply to the image display, the differences between the open and closed samples must be obtained.

If we consider one element of the array, modulated by square wave chopping, the output for that element under stable conditions will approximate to a triangular wave (see Figure 3.2), as shown schematically in Figure 6.15 A. This is sampled in successive fields at points a, b, and c, and the signal for that element may be obtained from either $b - a$ or $b - c$. This process, applied to every element of the array, is known as image difference processing.

However, in real life, the conditions are rarely stable and there is likely to be a drift superimposed on the element output, due, for example, to ambient temperature change, moving targets, or other variations with time. The element output is therefore more likely to be of the form shown in Figure 6.15 B. Here we have added to the triangular wave of amplitude 10 units a steady drift of 2 units per cycle. If we now apply the simple image difference processing, we get the incorrect answers of either $b - a = 11$ or $b - c = 9$. In order to get the correct answer, we must use three-point image difference processing to give the average of these two answers, i.e.:

$$b - \frac{(a+c)}{2} = 10 \tag{6.2}$$

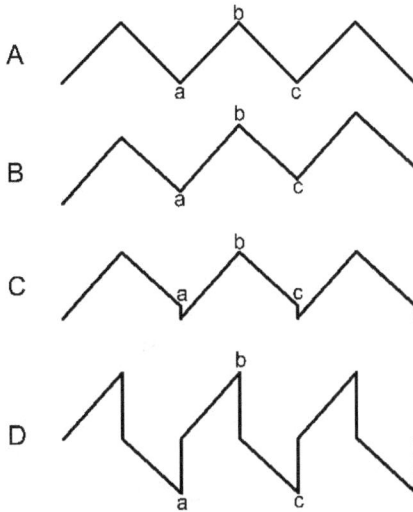

Figure 6.15 Schematic signal waveforms.

Another problem to address is that if we have a situation generating a constant drift, as shown in Figure 6.15 B, the array output will eventually saturate at one end or the other of its dynamic range. One solution to this problem is to use an electrically conducting pyroelectric material, as discussed in section 5.2.1.2. If the conductivity is chosen correctly, this will limit the low frequency response, so that slow temperature drifts are not a problem, but will not affect the pyroelectric response at the modulation frequency.

An alternative approach is to reset each element after sampling. If this is done once per frame (of two fields), a waveform as shown in Figure 6.15 C may be obtained. In this case, however, the differences are $b - a = b - c = 9$, and if the drift was in the opposite direction, the differences would be $b - a = b - c = 11$. So, we see that any drift either adds to or subtracts from the genuine signal.

If, however, we reset every field instead of once per frame, we get the waveform shown in Figure 6.15 D and the differences are $b - a = b - c = 20$, so we have only to divide this answer by 2 to get the correct value. As the differenced signal is doubled, this approach also has the potential advantage of increasing the signal to noise ratio of the system.

6.4.2.1. NETD & MRTD

The signal to noise ratio of a thermal imaging system is usually quoted in terms of the Noise Equivalent Temperature Difference (NETD). This is the temperature difference in a scene that produces an output signal equal to the noise.

Perhaps a more useful, but more subjective, parameter for a thermal imager is the Minimum Resolvable Temperature Difference (MRTD). This is the minimum temperature difference in a scene which can be seen by an operator viewing the display. This is subjective in that it depends on the skill of the operator, but it does give an indication of the performance of the whole system, including the display.

There are also trade-offs between thermal, spatial, and temporal resolution. For example, an imager operating at 8 Hz chopping frequency may have an NETD of 0.2 °C. The MRTD for a low spatial frequency scene (e.g. cool on the left side of the picture and warm on the right) may be 0.1 °C or better due to the temporal and spatial integrating effect of the display and the eye. This may be improved even further by integrating the signal over several frames, but, of course, this is at the expense of temporal frequency response. At high spatial frequency (one cycle per two array elements), however, the MRTD may be much worse than 0.2 °C.

6.4.3. Thermography

Thermography is the measurement of the temperature of an object remotely by means of its emitted infrared radiation. Single element pyroelectric detectors can be used to do this, but there is a problem knowing exactly what part and area of the scene is being measured. A thermal imager provides an excellent means for performing these measurements because the thermal image shows precisely the distribution of temperatures within the scene so that the user can see what is being measured.

The thermal imager shown in Figure 6.11 is, in fact, an imaging thermometer, and is indicating a temperature of 46.5 °C for the top of the radiator in the picture. The same device is shown in Figure 6.16, indicating that the temperature of the window above the radiator is 13.6 °C.

Figure 6.16 Imaging thermometer indicating the temperature of a window.

Great care must be taken in using a thermal imager to measure temperature. As explained in section 6.3.3, the instrument must be calibrated over the full range of ambient and target temperatures over which it is required to operate. There is also the problem of the emissivity of the item to be measured. If the object is completely black over the wavelength range of interest, i.e. its emissivity is 1, then the imager genuinely 'sees' the radiation emitted from the object and temperature measurement is relatively accurate. If, however, as is nearly always the case, the emissivity is significantly less than 1, then the imager views a combination of radiation emitted from the object and radiation from the surroundings reflected by the object. For example, rough concrete has an emissivity of approximately 0.94 so that 94 % of the observed radiation is emitted by the concrete and a fairly accurate temperature measurement can be made. Polished stainless steel, on the other hand, has an emissivity of only 0.075, so less than 10 % of the observed radiation is emitted by the steel, and any attempt to measure its temperature will, in fact, be measuring the temperature of other objects reflected in the steel surface. It is generally not recommended to attempt to use an infrared radiometer to measure the temperature of any object with an emissivity of less the 0.6, and even then, the operator needs to input the emissivity of the object and the temperature of the surroundings before making a measurement.

6.5.   Comparison with Alternative Detectors

The unique characteristic of pyroelectric detectors is their lack of DC response; that is a signal is produced only when there is a change in input radiation. This makes pyroelectric detectors ideal for sensing applications such as intruder alarms and people counting. The passive infrared intruder (PIR) alarm market is therefore dominated by pyroelectric detectors.

For applications where a DC or very low frequency response is required, there are other technologies which may offer better solutions. Some of these are briefly mentioned below.

6.5.1. Thermopiles

Thermocouples, consisting of two electrical junctions between two dissimilar electrical conductors, are frequently used for measuring temperature. If the two junctions are at different temperatures, a voltage is generated due to the Seebeck effect. A thermopile consists of several thermocouples connected in series (or possibly in parallel), with all the 'hot' junctions arranged to receive thermal radiation, and all the 'cold' junctions thermally connected to a heat sink. This arrangement can then be used as an infrared detector.

Thermopiles offer the advantages of a DC response and no microphony, but they have a much lower responsivity and signal to noise ratio than pyroelectric detectors at modulation frequencies above 1 Hz. A good comparison of pyroelectric and thermopile detectors is given by Neumann and Banta [40].

6.5.2. Resistive bolometers

A resistive bolometer consists of a resistor made of a material whose resistivity is highly temperature dependent, arranged so that its resistance can be monitored. Any change in temperature, caused, for example, by the absorption of infrared radiation, will be seen as a change in resistance. Much effort has been made by various bodies over the last few decades to develop two dimensional arrays of resistive bolometers for thermal imaging, and the performance of these arrays has now progressed to the point where they have almost universally replaced pyroelectric arrays for thermal imaging applications.

In order to measure the resistance of each individual bolometer element, a current must be passed through it and the voltage across it measured. Unfortunately, the current passing through the bolometer dissipates power and warms up the element. We thus have the situation that the act of measuring the resistance changes the temperature, and therefore the resistance. It is therefore much more difficult to read out the appropriate signals from a resistive bolometer array than it is for a pyroelectric. However, readout schemes have been developed that work very well, and the higher sensitivity and relative ease of manufacture of high element count arrays, has led to resistive bolometers dominating the uncooled thermal imaging market.

Because resistive bolometers have a DC response, no modulation of the incident radiation is required to produce an image. However, it is usually necessary to periodically re-calibrate the array to give a uniform image by interposing a shutter between the array and the imaging lens.

Kruse and Skatrud published a review of uncooled infrared imaging arrays and systems [41] that compares pyroelectric and resistive bolometer arrays in this application.

6.5.3. Photo-voltaic detectors

The highest performance thermal imaging systems are cooled systems using photovoltaic detectors, such as mercury cadmium telluride. These depend on the photovoltaic effect, in which free electrons are directly generated in a semiconductor material by the absorption of photons. Although these devices give the best performance, they do generally need to be cooled below ambient temperature, either by a thermoelectric cooler or by liquid nitrogen. For this reason, they tend to be more expensive than uncooled thermal imagers and have higher operating costs.

7. Infrared Optics

A complete infrared detection or imaging device consists of an infrared detector, with its associated electronics, together with infrared optics. The optical components need to do two things: to seal the package containing the detector, and to direct the infrared radiation onto the detector. It is possible to combine these two functions into one component, but it is normal practice to use a window to seal the package and a separate lens or mirror to focus the radiation. In this chapter we look very briefly at these components.

7.1. Infrared Windows

The calculations in sections 3.6 and 3.8 are for radiation falling on the detector surface. For all but a few exceptional circumstances, however, the detector package incorporates an infrared transmitting window. Without such a window, turbulence in the air would directly heat and cool the pyroelectric material, producing large pyroelectric noise, the input FET would be turned hard on by ambient visible light, and the high input impedance would be very susceptible to radio frequency pickup. We therefore need a window that protects the pyroelectric from air movement, cuts out visible light, and provides electrical screening; but transmits infrared radiation. The most common solution is to use a germanium window mounted into a metal housing with electrically conducting epoxy, as illustrated in Figure 3.7.

7.1.1. Germanium Windows

Germanium has low optical absorption for wavelengths from 2 μm to 15 μm, but it has a refractive index of about 4 in the infrared region, so it reflects 50% of the radiation incident upon it. Thin film optical coatings are therefore used to improve the infrared transmission. Following years of development, it is now possible to procure germanium windows with multilayer coatings (with as many as 17 layers on each surface) giving transmission of greater than 95% from 7.5 μm to 11.5 μm. The transmission spectrum of a window of this type is shown in Figure 7.1.

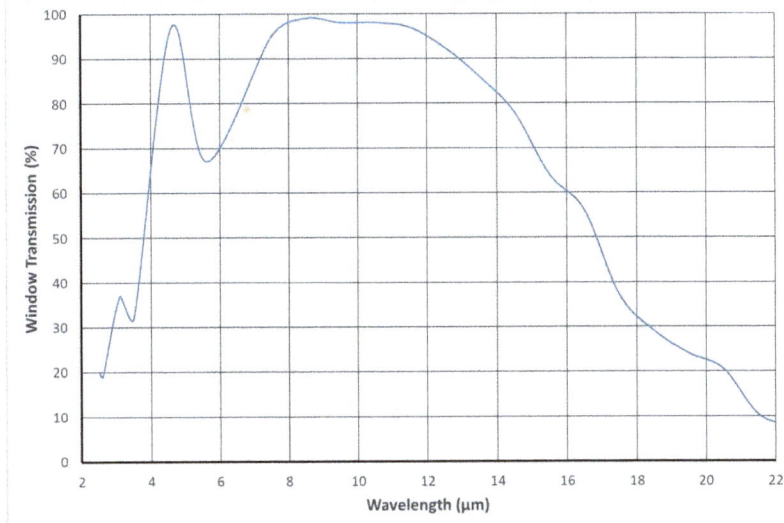

Figure 7.1 Transmission spectrum for high performance 8-12 μm window

This window will transmit 74% of the incident radiation from a black body at 20 °C and gives the best possible performance for detector arrays used in applications such as thermal imaging. For intruder detectors and people counters, however, there is an additional problem. As Figure 3.10 illustrates, there is significant atmospheric absorption at wavelengths between 5.5 μm and 7.5 μm. This absorption is mainly due to water vapour in the atmosphere. Not only does the water vapour absorb at these wavelengths, it also emits black body radiation, and if there is turbulence in the air this radiation will be modulated and will be seen by the detector and produce false alarms. For this reason, long-wave-pass and band-pass filters have been developed, as illustrated in Figure 7.2 and Figure 7.3.

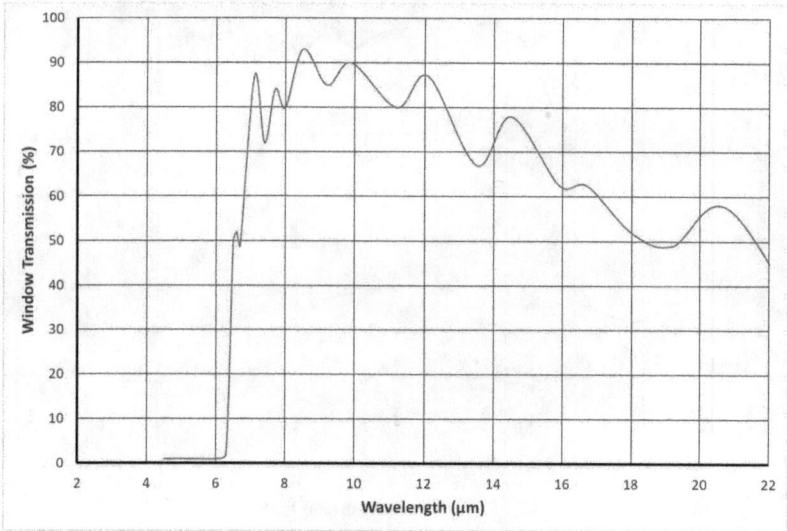

Figure 7.2 Transmission spectrum for a long wave pass filter.

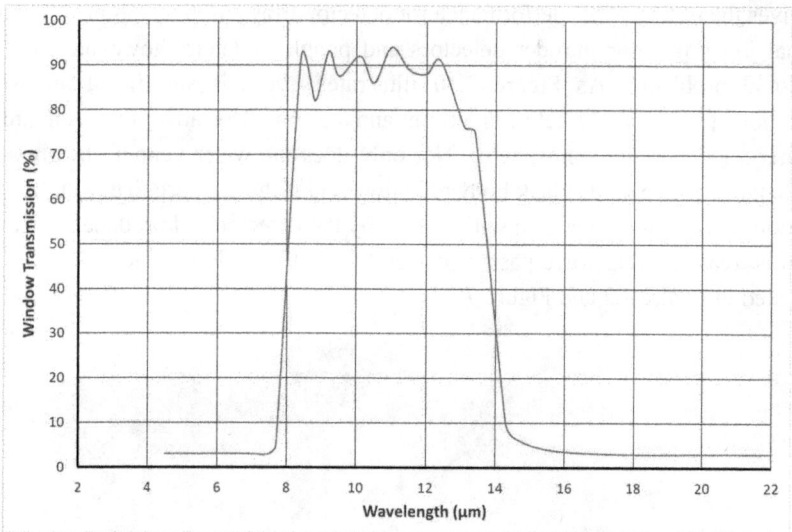

Figure 7.3 Transmission spectrum for a band pass filter.

7.1.2. Silicon Windows

Silicon also transmits infrared radiation and is much cheaper than germanium, but it has significant absorption bands at wavelengths longer than 6.5 µm. Optical grade silicon, with a low oxygen content, is manufactured specifically for infrared windows, and in thin sections this can give fairly good transmission up to 10 µm. Because of its low cost, silicon, with a suitable coating, is used for some people detectors. Silicon can also be used for applications such as flame detectors (see section 6.2.4) where a band pass filter centred at 4.3 µm is required.

7.1.3. Other Window Materials

There is a wide range of other materials that transmit infrared radiation [42]. These include alumina (which transmits from 0.3 µm to 3 µm), diamond (6 µm to beyond 300 µm), and silica (0.3 µm to 2 µm). Water soluble materials such as sodium chloride and potassium chloride (both 0.3 µm to 10.5 µm) have even been used in specialist applications. Many of these materials also transmit visible light so precautions must be taken to prevent visible light reaching the input transistor and amplifier, which are light sensitive.

7.2.   Infrared Lenses

7.2.1. Plastic Lenses

A thin film of polyethylene transmits radiation in the 8 µm to 14 µm wavelength range. Plastic is generally low cost and can be moulded easily, so most PIR intruder alarms and occupancy sensors incorporate a multi-faceted plastic lens similar to those illustrated in Figure 6.3 and Figure 6.7.

7.2.2. Germanium Lenses

For many applications germanium lenses are used. For people counters using a 16x16 element array, as in section 6.2.3, a single element spherical lens is adequate. As the array elements are on a 100 µm pitch and high image quality is not required, there is no point in paying for a more sophisticated lens giving better spatial resolution. For low-cost thermal imaging, a single element aspheric germanium lens may be used. For higher performance imaging applications, it may be necessary to use multi-element germanium lenses, as described, for example, by Riedl [43]. In order to get as much radiation as possible on to the detector, it is common practice to use lenses with an optical aperture of f/1, or even larger.

All these germanium lenses require antireflection coatings on all surfaces. For the inside surfaces of the lens assembly a high transmission coating with a transmission spectrum like that shown in Figure 7.1 may be applied, but on the front surface of the lens a more durable, though slightly less transmissive, coating is usually used. For extreme environments a diamond-like hard carbon coating is sometimes used.

7.2.3. Other Lens Materials

As mentioned in section 7.1.3 when discussing windows, there is a range of infrared transmitting materials that could be used. Silicon is sometimes used instead of germanium because of its lower cost. At the other extreme, where cost is less important, sapphire lenses are used in some flame detectors for fire alarms.

7.3.   Mirrors and Distorting Optics

Spherical or parabolic mirrors, made, for example, of aluminium, may be used instead of lenses. These are particularly suitable for long focal length optics, where the size and cost of germanium lenses would be prohibitive, but are of little benefit for short focal lengths.

The aim of most optical design work is to get the most accurate image of the object plane onto the image sensor. Mirrors can, however, be used to deliberately distort the field of view. If, for example, a pyroelectric array were to be used instead of a single element detector in an intruder alarm as described in section 6.2.1, the mounting arrangement shown in Figure 6.4 would mean that elements of the array viewing the far field would cover a much larger area of the ground than those viewing the near field. To even out these fields of view it is possible to interpose a specially contoured mirror between the lens and the object plane.

8. Conclusion

Having been first mentioned as an interesting phenomenon by an ancient Greek philosopher and studied by scientists in the nineteenth and early twentieth centuries, it was in the second half of the twentieth century that pyroelectricity became a subject of significant research and development and began to find major applications. Pyroelectric infrared detectors were first manufactured and deployed as person detectors in the 1970s, and the Passive Infrared Intruder (PIR) industry was established and experienced worldwide growth over the following four decades.

Research and development over the last fifty years has included, among other things, pyroelectric materials, detector and packaging design, development of amplification and signal processing electronics, software development, infrared optics design, and overall packaging development.

Significant research and development was aimed at producing thermal imaging systems. Initially this did result in the production of low-cost imaging systems, but this provoked efforts to reduce the cost of competing technologies that eventually matched the price of the pyroelectric versions and outperformed them. The requirement for continuous modulation of the infrared radiation is a significant disadvantage of pyroelectrics for thermal imaging. However, pyroelectric detectors continue to be the preferred technology for monitoring situations where a change in temperature within the scene occurs over a time period in the range 0.01 to 10 seconds.

Today, there are probably few new houses built in the western world that do not have at least one pyroelectric detector in a light switch or an intruder alarm. According to one report [44], the global motion sensor market was valued at USD 5.2 billion in 2018. About 50 % of this product contained pyroelectric detectors. Add to this other applications, including flame detectors, spectral analysis, and other instrumental applications, and we get some idea of the total size of the current market for pyroelectric detectors.

Who knows what future developments there may be? As the author once said when giving a presentation on the potential applications of pyroelectric detector arrays "The biggest market will probably be something we have not yet thought of".

9. References

[1]   S. B. Lang, Sourcebook of Pyroelectricty, New York: Gordon and Breach, 1974.

[2]   S. B. Lang, "A 2400 year history of pyroelectricty," *British Ceramic Transactions*, vol. 103, no. 2, pp. 65-70, 2004.

[3]   D. Brewster, *Edinburgh Journal of Science 1*, vol. 1, 1824.

[4]   J. Valasek, "Phys. Rev.," vol. 17, p. 475, 1921.

[5]   E. H. Putley, *Semiconductors and Semimetals*, vol. 5, p. 259, 1970.

[6]   E. H. Putley, *Semiconductors and Semimetals*, vol. 12, p. 441, 1977.

[7]   S. G. Porter, *Ferroelectrics*, vol. 33, pp. 193-206, 1981.

[8]   W. G. Cady, Piezoelectricity, vol. 699, New York: McGraw Hill, 1946.

[9]   J. C. Burfoot, Ferroelectrics, London: van Nostrand, 1967.

[10]   F. Jona and G. Shirane, Ferroelectric Crystals, New York: Macmillan, 1962.

[11]   G. S. Zhdanov, Crystal Physics, Edinburgh: Oliver and Boyd Ltd., 1965.

[12]   A. F. Devonshire, *Advances in Physics*, vol. 3, p. 87, 1954.

[13]   M. E. Lines and A. M. Glass, Principles and Applications of Ferroelectrics and Related Materials, Oxford: Clarendon Press, 1977.

[14]   P. Chandra and P. B. Littlewood, "A Landau Primer for Ferroelectrics," 5 February 2008. [Online]. Available: https://arxiv.org/abs/cond-mat/0609347. [Accessed 2020].

[15]   U.S. Navy, "ELECTRO-OPTICS," 13 September 2001. [Online]. Available: http://web.archive.org/web/20010913091738/http://ewhdbks.mugu.navy.mil/EO-IR.htm#transmission. [Accessed 30 July 2021].

[16]   T. V. Teperik, V. V. Popov and F. J. Garcia de Abajo, "Total light absorption in plasmonic nanostructures," *Journal of Optics A: Pure and Applied Optics*, vol. 9, pp. S458-S462, 2007.

[17]   B. I. Bleaney and B. Bleaney, Electricity and Magnetism, Oxford: ClarendonPress, 1965.

[18]   S. G. Porter, D. Appleby and F. W. Ainger, "Pyroelectric Ceramics for Thermal Imaging," *Ferroelectrics*, vol. 11, pp. 351-354, 1976.

[19]   R. C. Jones, *Advan. Electron.*, vol. 5, pp. 1-96, 1953.

[20]   J. B. Johnson, *Phys. Rev.*, vol. 32, pp. 97-109, 1928.

[21]   K. Kandiah, M. O. Deighton and F. B. Whiting, "A physical model for random telegraph signal currents in semiconductor devices," *J. Appl. Phys.*, vol. 66, p. 937, 1989.

[22]   R. C. Jones, "Quantum efficiency of photoconductors," *Proc. IRIS*, vol. 2, p. 9, 1957.

[23]   R. C. Jones, "Proposal of the detectivity D** for detectors limited by radiation noise," *J. Opt. Soc. Am.*, vol. 50, p. 1058, 1960.

[24]   Landolt-Bornstein New Series Group III, Ferro- and Antiferroelectric Substances, Berlin: Springer-Verlag, 1975.

[25]   H. P. Beerman, *Ferroelectrics*, vol. 2, p. 123, 1971.

[26]   A. M. Glass, *Phys. Rev.*, vol. 172, p. 564, 1968.

[27]   C. B. Roundy and R. L. Byer, *J. Appl. Phys.*, vol. 44, p. 929, 1973.

[28]   A. M. Glass, *J. Appl. Phys.*, vol. 40, p. 4699, 1969.

[29]   R. J. Phelan Jr., R. J. Mahler and A. R. Cook, *Appl. Phys. Lett.*, vol. 12, p. 57, 1972.

[30]   P. Buchman, *Ferroelectrics*, vol. 5, p. 39, 1973.

[31]   R. J. Mahler, R. J. Phelan Jr. and A. R. Cook, *Infrared Phys.*, vol. 12, p. 57, 1972.

[32]   B. J. Lillicrap, J. D. C. Wood, V. M. Wood and N. Shaw, "Growth and electrical properties of deuterated TGS produced by the rotating disc technique," *J. Phys. D: Appl. Phys.*, vol. 12, pp. 633-644, 1979.

[33]   E. T. Keve, K. L. Bye, P. W. Whipps and A. D. Annis, "Structural Inhibition of Ferroelectric Switching in Triglycine Sulphate - 1. Additives," *Ferroelectrics*, vol. 3, pp. 39-48, 1972.

[34]   G. H. Heartling and C. E. Land, *J. Amer. Ceram. Soc.*, vol. 54, pp. 1-11, 1971.

[35]   "Improvements in or relating to ceramic materials". United Kingdom Patent 1,514,472, 14 June 1978.

[36]   P. C. Osbond, C. S. Oswald and R. W. Whatmore, "Improvements relating to pyroelectric ceramics". Europe Patent 194,900, 17 September 1986.

[37]   G. Baker and A. D. Annis, "Pyroelectric detector circuits and devices". United States Patent 4,198,564, 15 April 1980.

[38]   R. Watton, D. Burgess and B. Harper, "The Pyroelectric Vidicon: A New Technique in Thermography and Thermal Imaging," *Journal of Applied Science and Engineering A,* vol. 2, pp. 47-63, 1977.

[39]   J. Nash and R. Saunders, "A review of Stratospheric Sounding Unit radiance observations in support of climate trends investigations and reanalysis," The Met Office, UK, 2013.

[40]   N. Neumann and V. Banta, "Comparison of Pyroelectric Detectors and Thermopiles," in *Proc. AMA Conferences IRS 2013*, Nurnberg, 2013.

[41]   P. W. Kruse and D. D. Skatrud, Uncooled Infrared Imaging Arrays and Systems Semiconductors and Semimetals Volume 47, San Diego: Academic Press, 1997.

[42]   A. J. Moses, Handbook of Electronic Materials Volume 1 Optical Materials properties, New York: Springer, 1971.

[43]   M. J. Riedl, Optical Design Applying the Fundamentals, Bellingham: SPIE, 2009

[44]   G. V. Research, "GVR-4-68038-055-2," November 2019. [Online]. Available https://www.grandviewresearch.com/industry-analysis/motion-sensors-market. [Accessed 30 July 2021].

Acknowledgements

I would like to thank Professor Roger Whatmore for inspiring me to write this book, for assistance in getting started, and for constructive criticism of early manuscripts. I would also like to thank Dr. Chris Carter for providing me with useful material and information, for reading the draft manuscript, and for making valuable comments and suggestions. Thanks also to all at Infrared Integrated Systems, both past and present, who were helpful and inspiring colleagues during my time there.

Finally thank you to my wife, Margery, for encouraging me to get the book finished.

About the Author

After gaining his BSc and PhD in Physics at the University of Leeds, Stephen Porter started work for the Plessey Company at the Allen Clark Research Centre in 1971. At Plessey (and subsequently GEC) he was engaged in research, development, and production of pyroelectric infrared detectors and arrays for a range of detection and imaging applications. This included leading an MOD funded programme to develop high-performance two-dimensional arrays for thermal imaging.

In 1996 Steve left GEC in order to join the newly established InfraRed Integrated Systems Ltd. (Irisys). Here, as Chief Engineer, he used the expertise gained at Plessey/GEC, but applied it in a cost reduced manner to the design and manufacture of low cost two dimensional pyroelectric arrays for intelligent detection and monitoring applications. The result was a 16x16 element pyroelectric array that became the heart of a range of thermal detection and imaging products. Over almost twenty years that he worked for Irisys, Steve made significant contributions to the design and development of a range of pyroelectric arrays, application specific integrated circuits, optics, and systems.

Steve has now retired and lives in England.